X-RAY INTERPRETATION FOR THE MRCP

P.S. Parfrey BSc MRCP
Lecturer in Medicine
Charing Cross Hospital
London

B.C. Cramer MRCP FRCR
Senior Registrar in Radiology
Northwick Park Hospital
London

CHURCHILL LIVINGSTONE
EDINBURGH LONDON MELBOURNE AND NEW YORK 1983

CHURCHILL LIVINGSTONE
Medical Division of Longman Group UK Limited

Distributed in the United States of America by
Churchill Livingstone Inc., 1560 Broadway, New
York, N.Y. 10036, and by associated companies,
branches and representatives throughout the
world.

© Longman Group Limited 1983

All rights reserved. No part of this publication may
be reproduced, stored in a retrieval system, or
transmitted in any form or by any means,
electronic, mechanical, photocopying, recording
or otherwise, without the prior permission of the
publishers (Churchill Livingstone, Robert
Stevenson House, 1-3 Baxter's Place, Leith Walk,
Edinburgh EH1 3AF).

First published 1983
  Reprinted 1985
  Reprinted 1988
  Reprinted 1989

ISBN 0 443 02594 0

British Library Cataloguing in Publication Data
Parfrey, P.S.
  X-ray interpretation for the MRCP.
  1. Diagnosis, Radioscope   2. X-rays
  I. Title   II. Cramer, B.C.
  616.07'572   RC78

Library of Congress Cataloging in Publication Data
Parfrey, P.S.
  X-ray interpretation for the MRCP.
  Bibliography: p.
  Includes index.
  1. Diagnosis, Radioscopic — Atlases.   2. Diagnosis,
Radioscopic — Examinations, questions, etc.
  I. Cramer, B.C.   II. Title. [DNLM: 1. Radiography —
Examination question WN 18 P229x]
  RC78.P27   1983   616.07'572'076   82-22071

Produced by Longman Group (FE) Ltd
Printed in Hong Kong

PREFACE

Every physician should be able to read accurately chest, plain abdominal, hand and skull X-rays. In addition, a knowledge of the radiological appearances of common medical conditions using modern imaging techniques is also necessary for the aspiring physician. Recently in postgraduate medical examinations, slides of radiographs have been introduced to test interpretation of these radiological appearances. The radiographs, questions and short answers presented in this book are similar to tutorials we have given to candidates for postgraduate examinations. However, in addition to being an aid for examination, we hope the reader will learn how to read radiographs in a logical sequence as suggested in the section entitled 'How to read radiographs', and that he will be able to review various radiological appearances by use of the subject index. We believe that heavily illustrated books similar to this one, have an important role to play in medical education as companions to the major textbooks, because they permit self examination and also illustrate many of the radiological appearances described in the textbook.

Five papers, each containing 20 questions with one to three stems are presented. Sometimes a clue is given with the question to help interpret the radiograph, especially when more than one correct interpretation could be made of the slide. Some of the questions we ask in this book are a little more difficult than those asked in the Membership of the Royal College of Physicians' diploma.

The book contains 164 radiographs which comprise 66 chest X-rays, 13 of which relate to cardiological disease but the vast majority of which relate to pulmonary disease, 12 plain abdominal X-rays, 6 skull X-rays, 9 hand X-rays and 16 X-rays of other bones. In addition there are 7 barium swallows, 7 barium meals/follow through, 6 barium enemas, 8 intravenous urograms, 7 arteriograms and 9 computed tomographs, most of which are head scans. A number of X-rays from more specialized procedures are also demonstrated, such as retrograde ureterogram, chest tomograph, ultra-sound, myelogram, venogram, percutaneous transhepatic cholangiogram and ERCP. Answers with a relevent discussion are given at the end of each paper. The amount of information given in the answers is not uniform or comprehensive. Readers should look up doubtful points in a textbook where they are treated more fully.

1983 P.S.P.
B.C.C.

ACKNOWLEDGEMENTS

This book has been written as a result of our experience in teaching MRCP candidates in Charing Cross and Northwick Park Hospitals where the vast majority of the radiographs originated. The quality of the reproductions in this book are a tribute to the Department of Medical Illustration, Charing Cross Hospital, and to Mr A.R. Williams, Head of the Department and his associates, we offer our thanks.

We also thank Dr A.W. Seed and Dr D. Katz for permission to use their radiographs, and Mrs Betty Spector and Miss Sheila McCoy for precise and generous secretarial assistance.

CONTENTS

READING RADIOGRAPHS 1

PAPER 1	*Questions*	7
	Answers	35
PAPER 2	*Questions*	49
	Answers	75
PAPER 3	*Questions*	97
	Answers	125
PAPER 4	*Questions*	143
	Answers	169
PAPER 5	*Questions*	191
	Answers	217

REFERENCES 235

INDEX 237

HOW TO READ RADIOGRAPHS

The chest film

Technical factors may alter radiographic appearances. Therefore one must observe whether the chest film is:
   (a) labelled as to side, name and date;
   (b) P-A or A-P; erect or supine;
   (c) of the correct penetration (the posterior ribs should be just visualized through the heart shadow);
   (d) taken following full inspiration;
   (e) rotated (rotation of the patient may be checked by comparing the medial ends of the clavicles on each side with the thoracic spinous processes).

Complete examination of the chest X-ray should include observation of:
   1. Soft tissues.
      e.g. mastectomy (11), subcutaneous emphysema (24), calcification (71).
   2. Bones of thorax, cervical spine and upper shoulders.
      e.g. fracture (119), metastases (211), collapsed vertebra (59), Looser zone, joint erosion, cervical rib (197), rib notching (57), absent clavicles (70).
   3. Diaphragm.
      In full inspiration the right hemidiaphragm lies between the anterior ends of the fifth and seventh ribs and the left hemidiaphragm is below the right.
   4. Trachea.
      If the trachea is not central it may be pushed across by a superior mediastinal mass (retrosternal goitre, 208) or pulled over to the side of the lesion by fibrosis or decrease in lung volume (154). It is rare for the trachea to be displaced by a pleural effusion (123) or pneumothorex, however large it may be.
   5. Superior mediastinum.
      e.g. retrosternal goitre (208), lymphadenopathy (104, 205),

Aneurysm (23), thymoma, dermoid cyst, neurogenic tumour (57), oesophageal enlargement.
6. Aorta.
   e.g. aneurysm (16, 17), dilation (149), coarctation, calcification.
7. Hilar regions.
   e.g. lymphadenopathy (20), pulmonary vessel enlargement (56, 116).
8. Cardiac size and silhouettte.
   The usual cardio-thoracic ratio i.e. the ratio of the maximum transverse diameter of the heart to the maximum transverse diameter of the chest taking the end points at the edge of the internal chest wall, is less than 1:2.
9. Lung fields.
10. Peripheral vessels.
11. Hidden areas.
    The heart, diaphragm and clavicles may obscure lesions which on careful examination can be detected.

The plain abdominal X-ray

When looking at a plain abdominal film observe
1. The name and date.
2. Erect or supine, inspiration or expiration?
   The gas pattern is best seen on a supine film but demonstration of free peritoneal air or fluid levels requires a horizontal film. Localization of opacities to a particular organ usually necessitates inspiration and expiration films, and perhaps oblique views also.
3. The gas pattern and position.
   Valvulae conniventes of small bowel reach from one wall to the other and the gas, when visible, is in the centre of the abdomen. Haustral folds of large bowel go only part of the way across the colon and the gas shadow is peripheral (110). Gas may be abnormal in position when pushed upwards by an ovarian cyst (53), or into the right lower quadrant by an enlarged spleen (26) or centrally by ascites (108).
4. Extraluminal gas.
   e.g. in the peritoneum (202) retroperitoneal space (205), the liver or biliary system (50), the genito-urinary system, colonic wall or subphrenic abscess.
5. Opacities in abdomen or pelvis.
   e.g. calcification in costal cartilage, arteries, lymph nodes,

(114), adrenals (205), pancreas (206), Liver or spleen (67), uterus (120), prostate, bladder; gallstones (round, faceted, single or multiple); renal or ureteric stones; parasites in abdominal wall; dermoid cyst with teeth (120).
6. Shape and size of solid organs.
7. Bones of spine and pelvis.
   e.g. osteolytic or osteoblastic metastases, Paget's disease, osteomalacia, avascular necrosis, osteoarthrosis.
8. Soft tissues, especially psoas shadows.

The hand X-ray

Every bone and joint must be looked at. Observe:
1. Deformities.
   e.g. syndactyly, polydaltyly, short fourth metacarpal, ulnar deviation.
2. Soft tissues.
   e.g. generalised increase in soft tissue thickness in acromgealy (171); localised thickness in gouty tophi (33); arterial, pericapsular or soft tissue calcification (163).
3. Joints.
   (a) Erosions
       In addition to rheumatoid arthritis and other similar connective tissue diseases erosive joint disease may occur in gout (33), psoriasis (111), Reiter's disease, osteoarthrosis (15), Charcot arthropathy.
   (b) Fluid in joints
       as suggested by widening of joint space and increased soft tissue shadowing around joint.
   (c) Distribution and symmetry of joint involvement.
4. Individual bone.
   (a) Bone density
       Diffuse loss of bone density may occur in osteoporosis, osteomalacia, hyperparathyroidism (102), infiltrative bone disease, whereas localised lucent areas may be seen in sarcoidosis (21), hyperparathyroidism (52), simple cyst, implantation dermoid, osteomyelitis and tumour. Coarse trabecular patterns occur in chronic haemolytic anaemia (74), Paget's disease and Gaucher's disease.
   (b) Erosions.
       Erosion of the terminal phalangeal tufts is an important sign of hyperparathyroidism (102) but may also occur in

scleroderma (200). An early sign of hyperparathyroidism is sub-periosteal erosions along the lateral margins of the middle phalanges (102).

The skull X-ray

In an adult skull X-ray observe:
1. Bone density and thickness e.g. diffuse increase of bone density in Paget's disease (211), osteoporosis, fluorosis; diffuse increase in bone thickness in acromegaly (144), chronic anaemia (96); localised increase in bone density in hyperostosis frontalis interna, meningioma fibrous dysplasia, tumours, osteomyelitis and Paget's disease, abnormal translucencies in congenital disease such as cleido cranial dysostosis (93), trauma, tumour (165) hyperparathyroidism (210), Paget's disease (211).
2. Position of normal and abnormal intracranial calcification. The calcified pineal is normally central, a deviation of more than 3 mm from the midline being abnormal.
3. The sella turcica. Normally the widest antero-posterior diameter of the pituitary fossa is 11–16 mm and the depth is 8–12 mm. Enlargement and ballooning of the sella turcica occurs with pituitary adenoma (144), erosion of the posterior clinoids with raised intracranial pressure, erosion of the lamina dura of dorsum sellae with tumour or aneurysm.
4. Air sinuses
5. Facial bones

Intravenous urogram

1. Control abdominal film. Look for calcification in renal areas (25, 114, 131, 207) ureters and bladder, for gas in urinary tract and at bone structure.
2. Size and position of kidneys. The normal length is 12–14 cm. The left kidney is normally between D12 and L2, the right being 1–2 cm lower than the left.
3. Shape and outline of kidneys e.g. multiple depressions in renal margins in pyelonephritis, ischaemic infarction or tuberculosis; localised bulge due to cyst, abscess, tumour (192), localised hydronephrosis or compensatory hypertrophy.
4. Renal substance thickness, which is the distance from

inter-calyceal line to renal margin (normally greater than 2 cm) e.g. generalised decrease in reflux or obstructive uropathy or renal artery stenosis; (215) localised decrease in pyelonephritis (with underlying calyceal abnormality) or ischaemia.
5. Nephrographic pattern e.g. nephrogram, which gradually becomes more dense in obstruction or severe underperfusion of kidney; immediate nephrogram which persists unchanged in acute tubular necrosis.
6. Shape and position of pelvi-calyceal system, e.g. (a) generalised rounding of the angles or clubbing of the calyces in hydronephrosis (21); (b) clubbing that is single or scattered in pyelonephritis, obstruction, tuberculosis or papillary necrosis (115); (c) clubbing with little or no loss of renal substance thickness in tuberculosis; (d) irregular or destroyed calyceal margins in tuberculosis, tumour (162), or papillary necrosis (115); (e) displacement of calyces in space occupying lesions such as cysts or tumour, and (f) filling defects caused by stones, tumour, blood clot or necrosed papillae.
7. Ureteric outline, e.g. enlargement due to obstruction (21, 102) or displacement towards midline as in retroperitoneal fibrosis, (68) or duplication (106).
8. Bladder shape, size and outline (102) both ante and post micturition.

Computed tomography

The basic assumption in computed tomography (CT) is that measurements taken of X-rays transmitted through the body contain information on all constituents of the body in the path of the beam. By multidirectional scanning multiple data is collected, mathematically assessed by computer and reproduced as horizontal cross sectional images.

Furthermore the demonstration of some structures is improved by the use of oral or intravenous iodinated contrast media. This is particularly evident in the head where marked enhancement following intravenous contrast may occur due to breakdown of the blood brain barrier or increased vascularity outside the brain. CT scanning is very valuable in the diagnosis of tumours (12, 152), abscess (199), haemorrhage (107) and infarction (22).

Fresh blood shows up as high attenuation areas. Therefore CT scanning is by far the most accurate radiological method of demonstrating haemorrhage, because of the clear distinction between the high attenuation of extravascated blood and that of surrounding brain. The cardinal sign of infarction is an area of decreased attenuation within the cerebral substance. Tumours cause variably reduced attenuation, but may compress the ventricular system. It is difficult to differentiate the tumour itself from surrounding oedema which is usually present. However, tumours frequently take up contrast, either patchily or circumferentially, making the contrast enhanced scan an important investigation in cases of suspected tumour.

PAPER 1

QUESTIONS

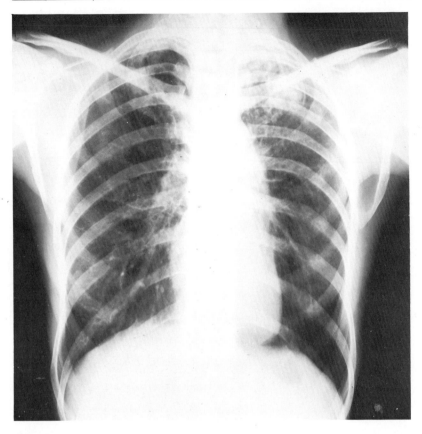

Q1.1

(a) What is the diagnosis?
(b) What serological test may confirm your diagnosis?

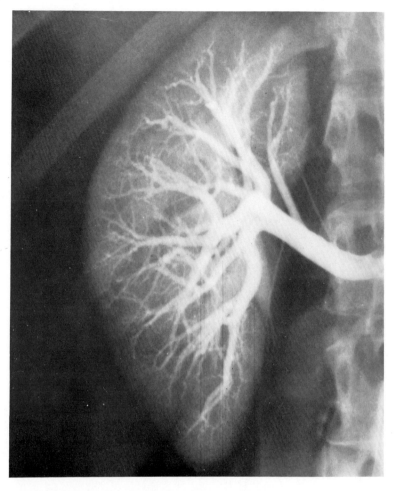

Q1.2 This patient presented with asthma.
  (a) What is the diagnosis?
  (b) Name four cutaneous manifestations of this disorder.

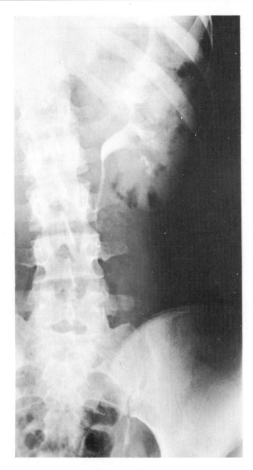

Q1.3 This patient has a right ureteric stone.
  (a) What diagnosis can be made from the appearance of the left kidney?
  (b) Name two other clinical manifestations of this disorder which may affect these patients.

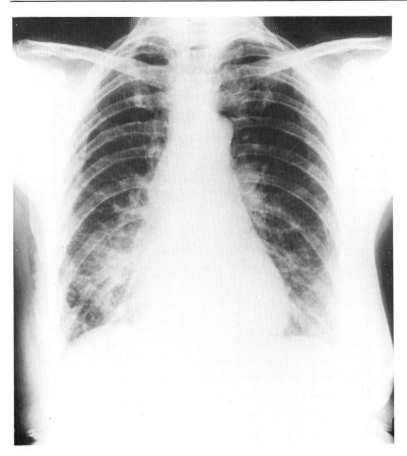

Q1.4 This 68 year old lady presented with dyspnoea.
  (a) Name 2 abnormal radiological signs.
  (b) What is the most likely cause of this lady's dyspnoea?
  (c) Name the characteristic abnormalities of pulmonary function in tests routinely available that you are likely to find in patients with this disorder.

ANSWERS PAGE (38)

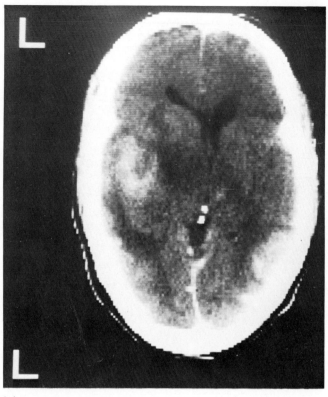

(1)

Q1.5 This 25 year old patient has congenital nerve deafness and presented as a 4 year old with failure to grow normally. One year prior to these investigations she complained of hoarseness.

(a) What diagnosis was made 20 years ago?
(b) What diagnosis was made 1 year ago?
(c) What is the cause of the radiological appearance in both X–rays?

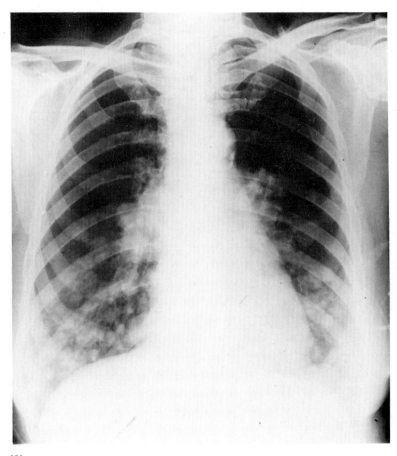

(2)

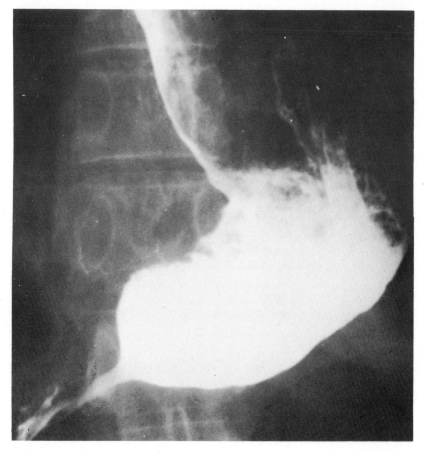

Q1.6

(a) Name the disorder demonstrated on this barium swallow.
(b) Name four ways in which this patient may present.

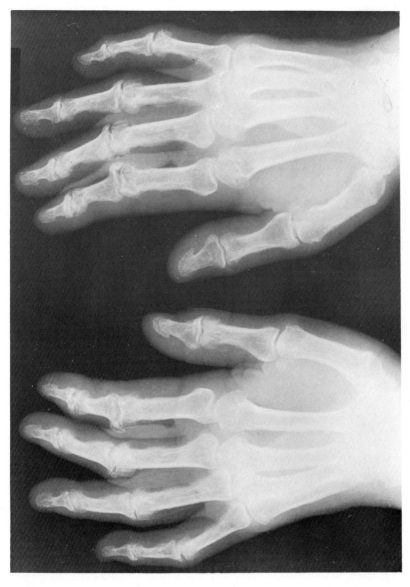

Q1.7

(a) What is the diagnosis?
(b) What other joints are most likely to be affected by this disease?

ANSWERS PAGE (40)

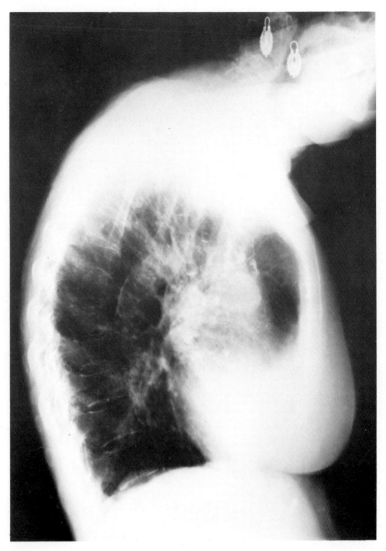

(1)

Q1.8 This patient presented with nocturnal angina unrelieved by glyceryl trinitrin.

(a) Name the abnormal radiological sign.
(b) What is the most likely cause of this abnormality?
(c) Name two other causes of this abnormality.

ANSWERS PAGE (40)

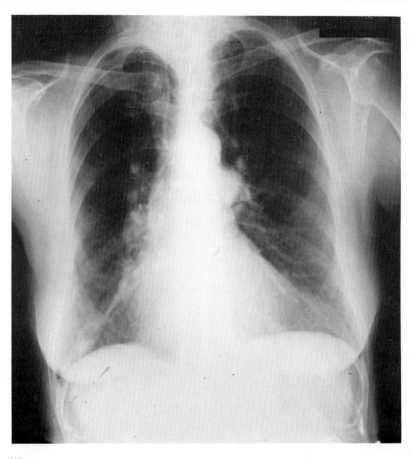

(2)

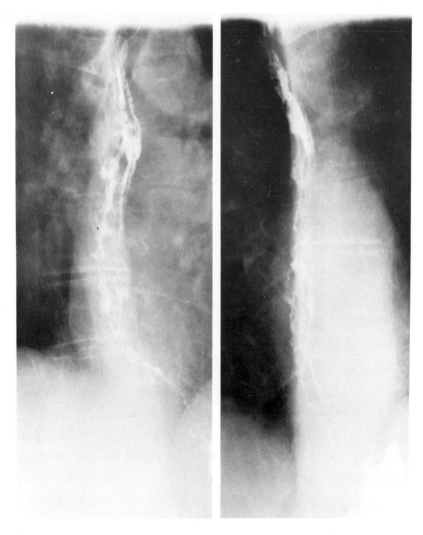

Q1.9

(a) Name the abnormal radiological sign.
(b) Name the four major clinical manifestations associated with this radiological appearance.
(c) Name the two frequent causes of this problem.

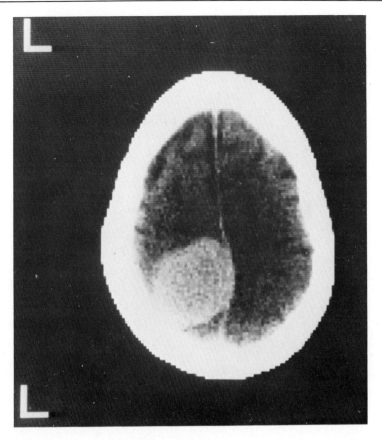

Q1.10 This is a post contrast CT scan.
  (a) What is the most likely diagnosis?
  (b) In what other sites does this lesion commonly occur — name three?

ANSWERS PAGE (41)

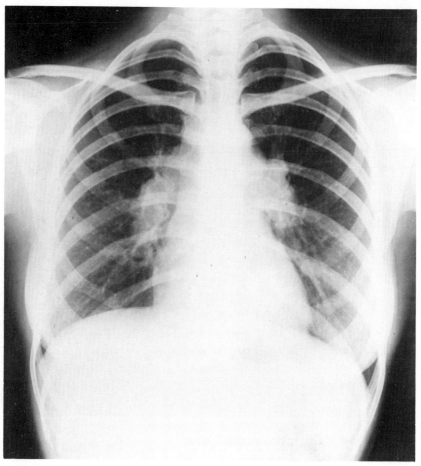

(1)

Q1.11 These three patients are suffering from the same disease.
  (a) What is it?
  (b) Name the radiological sign in each case.
  (c) What is the prognosis in patient 1?

ANSWERS PAGE (42)

PAPER 1 QUESTIONS

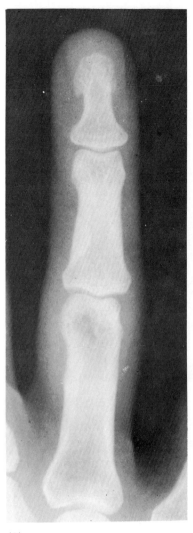

(2)

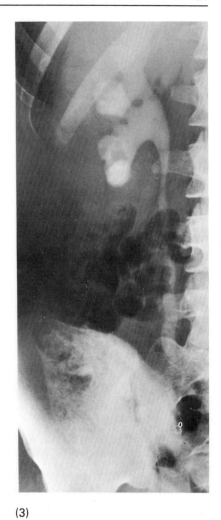

(3)

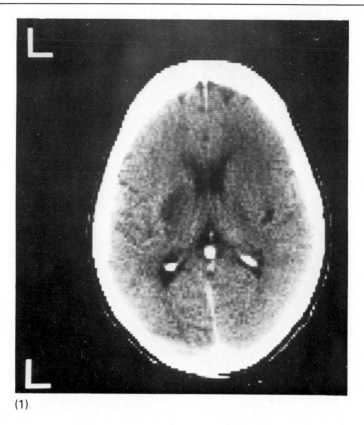
(1)

Q1.12 This 50 year old man was found collapsed in the street. On examination his left carotid pulse was absent.
  (a) What is the most likely cause of the abnormal radiological appearance in each X-ray?
  (b) Name two abnormal signs which may be heard on auscultation of this patient's heart.

ANSWERS PAGE (43)

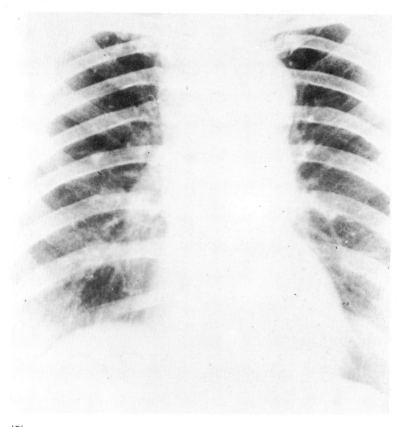

(2)

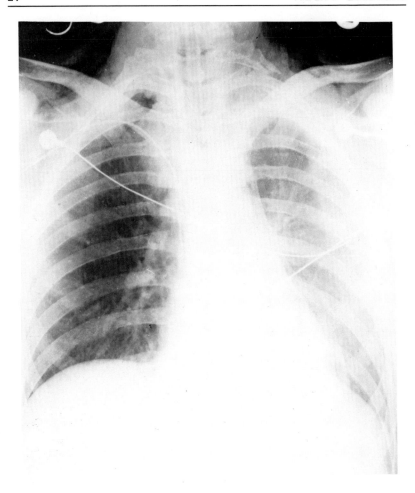

Q1.13

(a) Name four abnormal radiological signs.
(b) What is the most likely cause of these signs?
(c) Name the characteristic signs which you may hear in this patient.

ANSWERS PAGE (43)

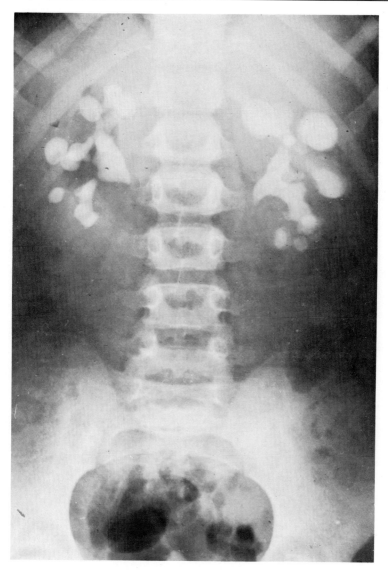

Q1.14

(a) What was the most likely initial chemical composition of these stones?

(b) Is the urine usually alkaline or acidic in patients who have these stones?

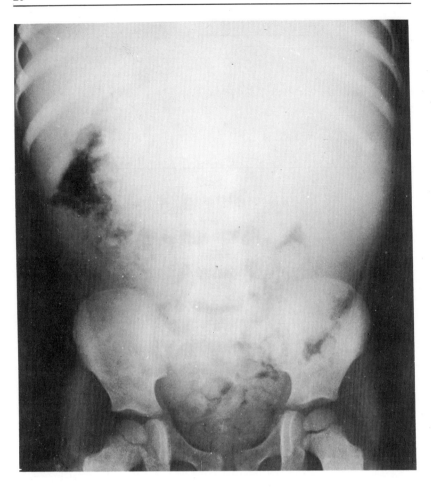

(1)

Q1.15 This patient is Jewish and presented at the age of two (abdominal film). Three years later he complained of left hip pain.
   (a) Name the abnormal radiological sign in both X-rays.
   (b) What is the most likely diagnosis?

ANSWERS PAGE (44)

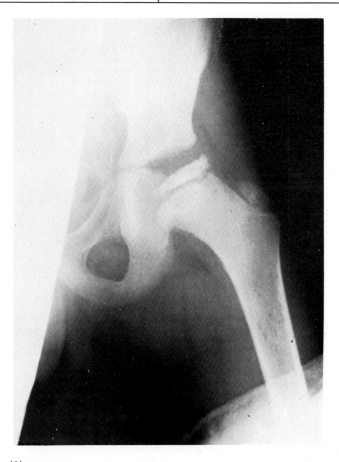

(2)

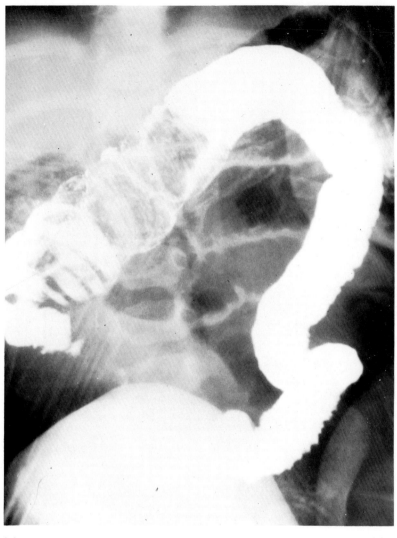

(1)

Q1.16 This patient is one year old.
  (a) What is the diagnosis?
  (b) What are the three initial symptoms which occur in this condition?

ANSWERS PAGE (45)

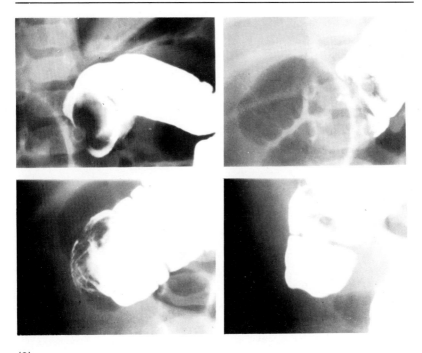

(2)

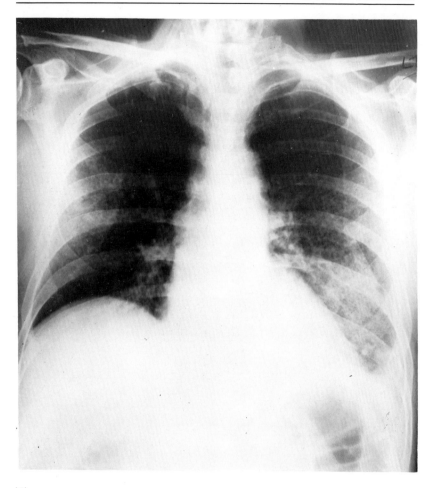

(1)

Q1.17 This patient presented with weight loss.
  (a) What is the most likely diagnosis?
  (b) How would you confirm your diagnosis?

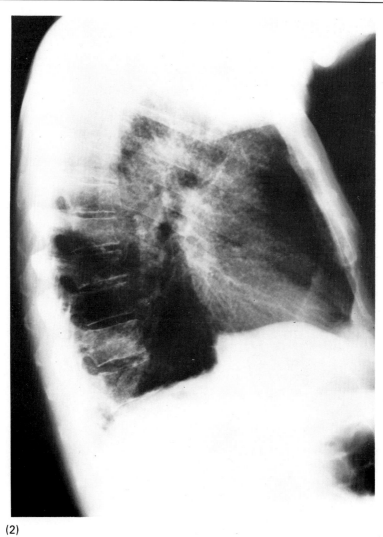

(2)

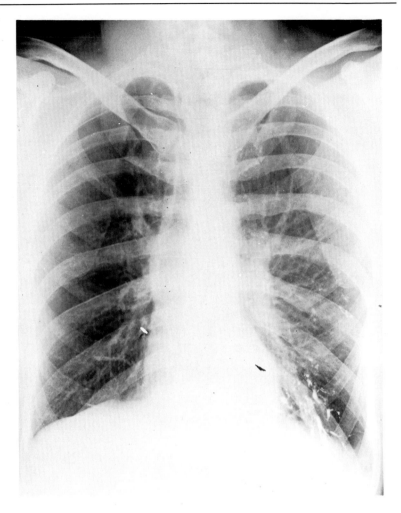

Q1.18 This patient is 20 years old.
  (a) What is the most likely cause of this radiological appearance?
  (b) What specialised radiological procedure had been performed prior to this X-ray
  (c) Name four contraindications to this procedure.

ANSWERS PAGE (47)

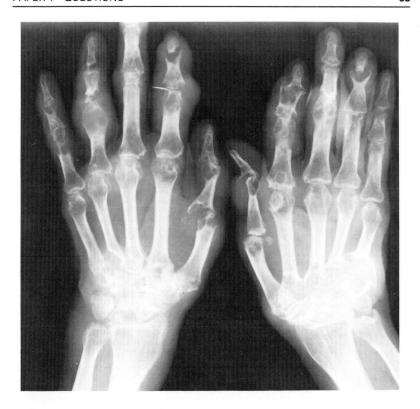

Q1.19

(a) What is the diagnosis?
(b) Name three forms of renal disease from which these patients suffer.

ANSWERS PAGE (47)

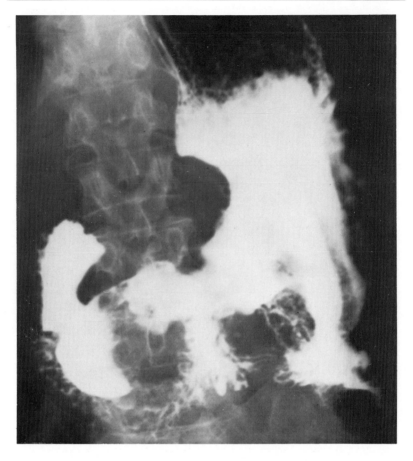

Q1.20

(a) Name four predisposing factors to this condition.
(b) If metastases occur in this patient how would they present — give four different presentations.

PAPER 1

ANSWERS

A1.1 (a) **Aspergilloma in a tuberculous cavity.**
(b) **Serum precipitins for aspergillus.**

The typical aspergilloma is a mass of mycelium, lying free within an open pulmonary cavity. There is usually no surrounding inflammation. The cavities in which an aspergilloma may develop are usually tuberculous but may be slowly resolving pneumonia, bronchial cysts, bronchiectasis, lung abscess, pulmonary infarction, pulmonary neoplasia or histoplasmosis. The development of the aspergilloma may be asymptomatic but mild haemoptysis is common. Weight loss, remittent fever and progressive deterioration of health may occur as the asperigilloma increases in size.

On the chest X-ray, an aspergilloma is recognized as a dense opacity separated from the wall of a cavity by a halo of air (Fig. A1.1). The opacity may move with different positions of the patient. Serum precipitins for aspergillus are almost always present in patients with aspergillomata.

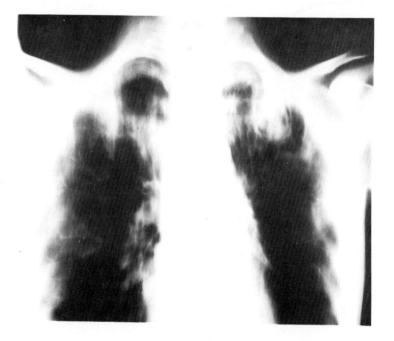

Fig. A1.1 A tomogram showing an aspergilloma in the upper zone of the left lung.

A1.2 (a) **Polyarteritis nodosa.**
    (b) 1. **Nonspecific rash;**
        2. **necrotizing vasculitis;**
        3. **livedo reticularis;**
        4. **nodules.**

Diagnosis of polyarteritis nodosa normally depends on biopsy. However arteriography may demonstrate aneurysms of medium-sized muscular arteries, as in this patient and help confirm the diagnosis.
Over a quarter of patients with polyarteritis nodosa develop cutaneous involvement in some form. In the acute stage of the disorder a multiform rash (urticarial or purpuric) or necrotizing vasculitis may occur. The latter may result in severe subcutaneous haemorrhage, secondary gangrene and ulceration. In the chronic stage of the disease livedo reticularis and ulceration occur. Cutaneous or subcutaneous nodules are unusual and may occur at any stage in the disease. The nodules may group, appear in crops, persist for periods of days to months, vary in size from 0.5 cm to 5 cm and cause the overlying skin to redden or ulcerate.

A1.3 (a) **Medullary sponge kidney.**
    (b) 1. **Urinary tract infection;**
        2. **haematuria.**

In medullary sponge kidney, the diagnosis is usually made on intravenous pyelography performed because of a urinary infection or haematuria. Its characteristic radiological signs are pools or streaks of contrast medium within the pyramids, without caliceal deformity, which sometimes appear to branch away from the calices like a bouquet of flowers. Most cases show calcification on the plain film, typically distributed as groups of small calculi in the pyramids. After injection, the calculi become surrounded by contrast medium and additional pools and/or streaks appear, which distinguishes medullary sponge kidney from other causes of nephrocalcinosis. Also, unlike nephrocalcinosis, medullary sponge kidney may affect one kidney alone or even a segment of one kidney. In papillary necrosis, there is usually only one large blob of contrast medium to a papilla, whereas in medullary sponge, the pools are small and numerous.

A1.4 (a) 1. Right mastectomy;
2. bilateral lower zone shadowing.
(b) Lymphatic carcinomatosis of breast.
(c) 1. Reduced forced vital capacity (FVC);
2. reduced forced expiratory volume (FEV) proportionate with FVC;
3. reduced PO_2 and normal PCO_2 of arterial blood;
4. reduced transfer factor for carbon monoxide (DCO).

Lymphatic carcinomatosis occurs when tumour invades the mediastinal glands and spreads along the lymphatics into both lungs. This is seen on chest X-ray as bilateral hilar enlargement with streaky shadows fanning out into both lung fields. The carcinoma may originate in organs other than lung, such as stomach, breast and pancreas. The symptoms of lymphatic carcinomatosis are progressive dyspnoea, cough and recurrent haemoptysis.

The functional pulmonary abnormality in these patients is initially a diffusion defect, which later develops into a restrictive lesion. The most useful tests of pulmonary function which are routinely available include FEV in one second (or peak expiratory flow rate), FVC, PaO_2, $PaCO_2$ and DCO. Just as the FEV_1 is the most useful monitor of obstructive lung disease so the FVC is the most useful monitor of restrictive lung disease. The characteristic pattern of disordered pulmonary function seen in patients with restrictive lung disease is reduced FVC and FEV_1 but normal FEV/FVC, reduced PO_2 and DCO but normal PCO_2.

A1.5 (a) Pendred's syndrome.
(b) Carcinoma of the thyroid.
(c) Metastases to the lung and brain.

Familial goitre and nerve deafness comprises Pendred's syndrome, and results from a deficiency of the peroxidase enzyme necessary to oxidize thyroidal iodide to iodine. Affected individuals have a goitre but are usually enthyroid or mildly hypothyroid. Continued TSH stimulation of the thyroid may occur in these individuals with congenital metabolic defects of thyroid hormone production, due to inadequate therapy with suppressive doses of thyroid hormone or potassium iodide. This produces hyperplasia, nodule formation, autonomous function and occasionally malignant change in the thyroid.

As many as 10 per cent of patients with anaplastic thyroid carcinoma have a 20 year history of goitre. Anaplastic carcinoma presents as a hard thyroid mass fixed to adjacent neck structures with symptoms such as cough, hoarseness, and dysphagia. Metastasis to cervical nodes and lungs occurs rapidly. The chest X-ray in this patient shows the trachea displaced to the right by a goitre and multiple nodular infiltrates in the lung fields. The CT brain scan with contrast shows a number of lesions surrounded by low attenuation lucent oedematous areas with distortion of the ventricular system, suggestive of multiple metastases.

A1.6 (a) Achalasia of the oesophagus.
  (b) 1. Dysphagia;
   2. substernal pain;
   3. weight loss;
   4. pneumonia or lung abscess.

Achalasia is a chronic disorder of oesophageal motility that leads to obstruction at the level of the lower oesophageal sphincter. It is due to degeneration of the ganglion cells of the myenteric plexus or of the vagal motor nuclei. The major symptom is dysphagia, which is usually insidious in onset. From the beginning, difficulty may be experienced in swallowing liquids as well as solids, although the patient may obtain relief by drinking fluids after a meal. Substernal pain after eating occasionally occurs in some patients, and lasts for a few minutes. In severe cases, loss of weight may be striking. Pulmonary problems, such as pneumonia and lung abscess, may also occur because of aspiration of oesophageal contents.

In the early stages the radiological appearances of achalasia are those of a dilated oesophagus with small irregular contractions of the oesophageal wall. With further dilation, the oesophagus becomes elongated and its lower end redundant. At the cardio-oesophageal orifice, the oesophagus ends in a 'cigar shaped' termination and considerable stasis of oesophageal contents may be seen. Absence of air from the gastric fundus is an additional feature.

A1.7 (a) Osteoarthrosis.
  **(b) 1. Cervical spine articulations between C4 and C5, C5 and C6;
    2. lumbar spine disc between L4 and L5, and L5 and S1;
    3. hips;
    4. knees.**

This patient's hand X-rays demonstrate marked joint space narrowing, osteophyte formation in the distal and proximal interphalangeal joints, cystic changes in subchondral bone, and absence of osteoporosis — all signs of osteoarthrosis. Degenerative joint disease of the wrists, elbows and shoulders is uncommon, whereas it is common in weight bearing joints such as the hips, knees and spine. In the cervical spine maximal motion occurs between C4 and C5, and between C5 and C6. Osteophytes develop from the margins of the joints of Luschka and the apophyseal joints, which may impinge on nerve roots at the spinal foramina. In the lumbar spine herniated disc disease is more common than impingement of nerve roots or cauda equina by osteophytes. Osteoarthrosis of the hip is the most disabling form of the disease with considerable loss of function in the advanced stages.

**A1.8 (a) Aneurysm of the ascending aorta.
  (b) Syphilis.
  (c) Arteriosclerosis and cystic medial necrosis.**

In the past nearly all aneurysms of the ascending aorta were due to syphilis but now the commonest causes are arteriosclerosis and cystic medial necrosis. The latter may occur in association with Marfan's syndrome, hypertension or ageing. In addition to aneurysms syphilis also gives rise to angina pectoris and aortic incompetence. Attacks of syphilitic angina tend to be longer than angina of atheromatous origin, are often nocturnal and glyceryl trinitrate may not be particularly effective. Radiologically, calcification confined to the ascending aorta (seen best in the lateral view as in Q1.7) is diagnostic of syphilis.

PAPER 1 ANSWERS 41

A1.9 (a) **Oesophageal varices.**
    (b) 1. **Haematemesis or melaena;**
        2. **splenomegaly;**
        3. **ascites;**
        4. **hepatic encephalopathy.**
    (c) 1. **Cirrhosis of the liver;**
        2. **mechanical obstruction of extrahepatic portal vein.**

The barium swallow in this patient shows widespread large cobblestone pattern suggestive of varices. Varices do not cause dysphagia or other oesophageal symptoms. They result from portal hypertension which induces collateral channels between the portal and systemic venous beds. In addition to the cardio-oesophageal junction, major sites of collateral flow include the rectum giving haemorrhoids, the retroperitoneal space giving dilated abdominal wall veins, and the falciform ligament of the liver, resulting in visible periumbilical veins (caput medusae).

    The commonest cause of portal hypertension is cirrhosis, around half the patients with cirrhosis having an important degree of portal hypertension. The second most common cause is mechanical obstruction of the extrahepatic portal vein, which usually results from thrombosis or tumour invasion. About 5 per cent of patients with cirrhosis and portal hypertension have associated portal vein thrombosis. Occlusion of the major hepatic veins (Budd-Chiari syndrome) or their small intrahepatic branches (veno-occlusive disease) may also lead to portal hypertension.

A1.10 (a) **Parasagittal meningioma.**
     (b) 1. **Sphenoid ridge and its convexities;**
        2. **olfactory groove;**
        3. **around sella turcica.**

The CT scan cut through the surface of the cerebral cortex, shows a high attenuation (opaque) lesion attached to the falx and extending across the posterior surface of the left cerebrum. The site and marked enhancement following contrast make a parasagittal meningioma, arising from the superior sagittal sinus, the most likely diagnosis.

    A tumour of the falx occurs frequently in the region of the paracentral lobule and is likely to produce weakness of both lower limbs beginning in the feet, one being usually affected more than the other. Retention of urine may occur owing to compression of the cortical centres for initiating micturition, and impairment of position sense in the toes may also be elicited.

Meningiomas arise from the cells of the arachnoid villi and consequently are found along the course of the intracranial venous sinuses. In addition to the superior sagittal sinus (parasagittal meningioma), the sites of greatest predilection are the sphenoparietal sinus and middle meningeal vessels (meningioma of sphenoid ridge and its convexities), the olfactory groove of the ethmoid, and the circle of sinuses around the sella turcica (suprasellar meningioma). Meningiomas are uncommon below the tentorium.

A1.11 (a) Sarcoidosis.
  (b) 1. Hilar lymphadenopathy;
  2. bone cyst;
  3. ureteral dilation to pelvic brim and hydronephrosis (secondary to a ureteral stone).
  (c) Spontaneous remission within 6 months to 2 years.

Sarcoidosis is a multisystem granulomatous disorder of unknown aetiology which most frequently presents with hilar lymphadenopathy, pulmonary infiltration, and skin or eye lesions. In the acute disorder of erythema nodosum and hilar lymphadenopathy remissions are frequent and occur over a period of 6 to 24 months. These patients have little or no evidence of residual disease and have normal life expectancy. Parenchymal lung involvement, which is seen in about half of all patients with sarcoidosis, worsens the prognosis and is the most frequent cause of death.

Asymptomatic punched out lesions in the phalanges of hands and feet are visible in X–rays of about 10 per cent of patients with sarcoidosis. These are usually associated with overlying skin lesions which vary from extensive erythematous, infiltrating and raised lesions to small nondescript plaques and papules. Both skin and bone lesions are associated with chronic sarcoidosis.

Increased sensitivity to vitamin D leads to hypercalcaemia and hypercalciuria, which may cause nephrocalcinosis or renal calculi. Impaired renal function may also occur secondary to hyperuricaemia and, rarely, because of direct granulomatous involvement.

PAPER 1 ANSWERS 43

A1.12 (a) 1. Left cerebral infarct;
2. dissecting aneurysm.
(b) 1. Aortic regurgitation;
2. pericardial friction rib.

The CT scan shows a low attenuation area, with no associated mass effect, in the region of the left internal capsule. In the first week after cerebral infarction about three quarters of patients have a low density area with variable ill defined or sometimes well defined margins on CT scan. The chest X-ray demonstrates a widened mediastinum extending to the right and the left which suggests dissection of the ascending aorta.

Since dissection of the ascending aorta involves the great vessels in the majority of cases, a discrepancy between the carotid pulses, or a difference in the blood pressure in the two arms should lead to a search for other evidence of aortic dissection. In this patient the dissection involved the left carotid artery and caused sudden right hemiplegia with diminished consciousness. In half of patients with ascending aorta dissection acute aortic regurgitation occurs. The appearance of a pericardial friction rib may be rapidly followed by pericardial tamponade due to haemopericardium.

A1.13 (a) 1. Fracture first left rib;
2. surgical emphysema;
3. mediastinal emphysema;
4. hazing of the left lung due to haemothorax.
(b) Trauma causing rupture of bronchus.
(c) Mediastinal crunch and Hamman's sign.

In this patient's chest X-ray mediastinal emphysema can be discerned by the double outline along the left heart border due to air between mediastinum and left lung. The air in the mediastinum may escape upwards into the subcutaneous tissues causing surgical emphysema of the neck and also, rarely, downwards into the retroperitoneal tissues where it may result in intestinal pneumatosis.

Mediastinal emphysema is seldom a primary event. Air enters the mediastinum directly from a ruptured bronchus or oesophagus; or indirectly along the perivascular sheaths of the pulmonary vessels, following rupture of lung alveoli; or through the retroperitoneal tissues, in the rare case in which mediastinal emphysema follows rupture of some part of the gastrointestinal tract.

Coarse crepitant sounds over the mediastinum (Hamman's sign) are sometimes best heard over the left sternal edge from the 3rd to the 6th intercostal space with the patient sitting up. The sounds are in time with the heart beat or, if present only in the upper sternal area, with respiration or with swallowing. Loud systolic crunching or clicking sounds may be heard over the precordium, especially if a pneumothorax is also present. Hamman's sign is not diagnostic of mediastinal emphysema.

A1.14 (a) Magnesium ammonium phosphate.
      (b) Alkaline.

The plain abdominal X-ray illustrates multiple large radio-opaque stones in the collecting system of both kidneys. Magnesium ammonium phosphate stones represent about one fifth of all urinary tract stones. They form in the presence of a urinary tract infection with an urease-producing organism which can split urea. This produces an alkaline urine and high urinary ammonium ion concentration. The urine alkalinity increases the relative concentration of the divalent phosphate and thus increases the ammonium phosphate ion product. Stones produced by urinary infection are relatively radio-lucent at first but eventually become strongly radio-opaque due to their increasing content of calcium phosphate. Initially they may be multicentric in origin but grow rapidly by creating further obstruction and infection. It should be remembered that urinary tract infections and consequent stone formation may occur in patients with a primary metabolic abnormality. However the stones of these patients with metabolic renal lithiasis are characteristically small, sharp, densely radio-opaque and formed from calcium oxalate and/or calcium phosphate.

A1.15(a) 1. Splenomegaly;
           2. avascular necrosis.
      (b) Gaucher's disease.

The abdominal X-ray in this patient shows that the air-filled bowel has been displaced down towards the right lower quadrant by a dense shadow, which is a huge spleen. The left hip is intact at this time but a few years later the head of the left femur has been flattened, suggestive of avascular necrosis.

The combination of massive splenomegaly and avascular necrosis in a Jewish child is strongly indicative of Gaucher's disease.

Gaucher's disease is a glycosylceramide lipidosis caused by deficiency of glucosylceramidase. It is an autosomal recessive disease and about 30 times more frequent in Ashkenazi Jews. An infantile form is characterized by marked hepatosplenomegaly, severe neurological disease and early death. Juvenile and adult forms occur with hepatosplenomegaly, hypersplenism, thrombocytopaenia, bone pain, pathological fractures, avascular necrosis and pulmonary involvement. The clinical course is variable. Pulmonary involvement may lead to early death but in other patients life span is not shortened.

A1.16 (a) **Intussusception.**
     (b) 1. **Abdominal pain;**
        2. **vomiting;**
        3. **bloody stools.**

About two thirds of patients who develop intussusception are under the age of 1 year. It is unusual to find a cause for intussusception in young children, but in older patients a polyp, neoplasm or other local lesion may be the initiating cause.

The initial symptom in almost half of cases is colicky abdominal pain and in the other half vomiting. In a few cases bloody rectal discharge is the first recognised symptom. Pain may decrease in intensity late in the course of intussusception when intestinal dilation produces atony. Vomiting appears initially as a reflex symptom but later on is the result of intestinal obstruction. The bloody stools consist of blood mixed with mucus to produce the classical currant jelly stool, but at times only a thin bloody fluid (prune juice stool) appears.

Barium enema is not only diagnostic of intussusception but may be therapeutic. Within 48 hours of initial symptoms the intussusception may be reduced during a barium enema by hydrostatic pressure. In this patient's barium enema the barium column is concave at the point of intussusception outlining it much as a column of barium in the vagina would outline the cervix.

A1.17 (a) Miliary tuberculosis.
 (b) Liver, lymph node or marrow biopsy.

Dissemination of tubercle bacilli throughout the body, following the release of the contents of a caseating lesion into a vein, results in miliary tuberculosis. The symptoms comprise weight loss, weakness, dyspnoea, fever, sweats and gastrointestinal disturbance. The white cell count may be normal, or low, or show a leukaemoid reaction or a monocytosis. The diagnosis is frequently not made until appearance of the typical miliary pattern on the chest X-ray. Histoplasmosis, blastomycosis, crytococcosis or miliary carcinomatosis may have similar clinical and radiological picture. Therefore it is vital to establish the correct diagnosis by biopsy in search of caseating granulomas and tubercle bacilli.

A1.18 (a) Bronchiectasis.
 (b) Bronchogram.
 (c) 1. Minimal disability;
  2. the presence of an acute exacerbation;
  3. generalized disease;
  4. chronic obstructive airways disease.

In saccular bronchiectasis multiple 1–2 cm cystic lesions or fluid levels may be seen on the chest X-ray. Usually, however, streaky infiltrate and loss of volume in bronchiectatic areas are visible. This patient had a bronchogram one week prior to this postero-anterior X-ray, which consequently revealed retention of contrast medium in the saccules of the left lower lobe.

The pathogenesis of bronchiectasis depends on factors which cause or lead to necrosis of the bronchial wall and supporting tissues. The origin of the destructive process is nearly always a bacterial infection but other factors, either hereditary, congenital or mechanical may also be present that predispose to the development of infection. In young adults with the disease infections have normally begun in childhood. As the number of recurrent bouts increase, the time between them tends to shorten and the response to treatment becomes less complete until chronic symptoms of cough and sputum develop. These patients are likely to have cystic fibrosis, immune-deficiency disease or atopic asthma.

In patients with suggestive symptoms bronchography is

indicated in the evaluation of patients for possible surgery, those with recurrent localized pneumonias or severe haemoptysis.

A1.19 (a) Chronic tophaceous gout.
(b) 1. Uric acid stones;
2. interstitial nephritis;
3. pyelonephritis;
4. glomerulosclerosis.

The characteristic tophus consists of deposits of urate crystals and intercrystalline matrix which elicit a low grade chronic inflammatory process with disruption of tissues and fibrosis. Destruction of tissue is particularly evident in cartilage and bone, leading to radiolucent punched out lesions of bone and gross deformities of joints. Tophi adjacent to the joints may contain calcific deposits and can usually be distinguished from joint effusions by their lobulated margins. Erosive lesions are usually periarticular and initially located beyond the capsule and synovial reflection, thereby distinguishing them from the marginal erosions of rheumatoid arthritis.

About one half of all patients with clinical gout develop visible tophi at some stage of their disease. Renal disease is the most frequent complication of gout except for the arthritis. The incidence of uric acid calculi in various series of patients with gout is between 5 and 35 per cent. Gouty nephropathy in the absence of arthritis, tophi and stones is very rare. Gouty nephropathy represents a variety of forms but the earliest structural abnormality is tubular damage associated with an interstitial reaction. The distinctive histological feature is the presence of urate crystals and surrounding giant cell reaction, which leads to tissue destruction, distortion and fibrosis. In addition to this primary lesion uric acid stones lead to hydronephrosis and increased susceptibility to chronic pyelonephritis. Also a distinctive glomerulosclerosis with thickening of glomerular capillary basement membrane is seen, which may produce proteinuria. End stage renal failure occurs in about a fifth of gouty subjects, following a slowly progressive decrease in glomerular filtration rate.

A1.20 (a) 1. Atrophic gastric mucosa;
2. pernicious anaemia;
3. adenomatous gastric polyps;
4. Billroth II partial gastrectomy.
(b) 1. Abdominal distension;
2. jaundice;
3. bone pain;
4. breathlessness;
5. neurological symptoms.

This barium meal demonstrates a large gastric carcinoma with a large filling defect in the antrum, irregular greater curve and shouldering. In patients with symptoms suggestive of gastric carcinoma an abnormality is demonstrated radiologically in about 90 per cent of cases, but in about a quarter of these patients it is impossible to differentiate benign from malignant gastric lesions. The radiological signs which favour a benign ulcer include projection beyond margin of lesser curve, radiating mucosal folds of normal width up to the ulcerated area, ulcer greater than 2 cm in diameter, and reduction in size after one month's medical treatment; whereas malignant ulcers are usually within margin of lesser curve, non-radiating thickened folds are interrupted near the ulcer, the ulcer diameter is usually one to two cm and there is little healing with medical treatment.

A small genetic element in the pathogenesis of stomach carcinoma is suggested by a greater incidence in relatives of patients with the disease and in people with blood group A. Atrophic gastric mucosa increases the risk of gastric carcinoma. This mucosal change occurs in pernicious anaemia, 6 per cent of whom develop stomach cancer. Adenomatous gastric polyps greater than 2 cm in diameter frequently contain adenocarcinoma. Ten to twenty years after a Billroth II partial gastrectomy for peptic ulcer there is an increased risk of gastric cancer.

The initial symptoms of patients with gastric carcinoma may be related to metastases. Abdominal distension may result from malignant ascites; jaundice from biliary tract obstruction by nodes in the porta hepatis or intrahepatic metastases; pain from bone secondaries; breathlessness from lung spread, and neurological symptoms from brain metastases.

PAPER 2
QUESTIONS

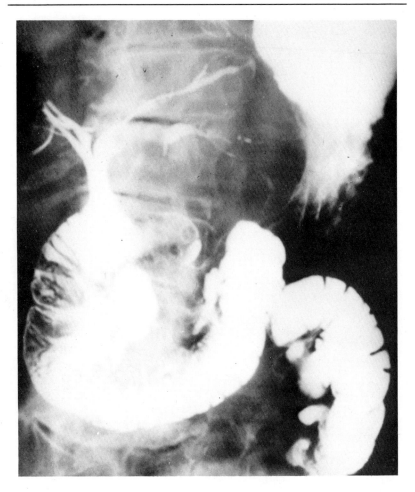

Q2.1 This 75 year old patient presented with intestinal obstruction.
   (a) Name two abnormal radiological signs.
   (b) What is the most likely cause of the obstruction?

ANSWERS PAGE (76)

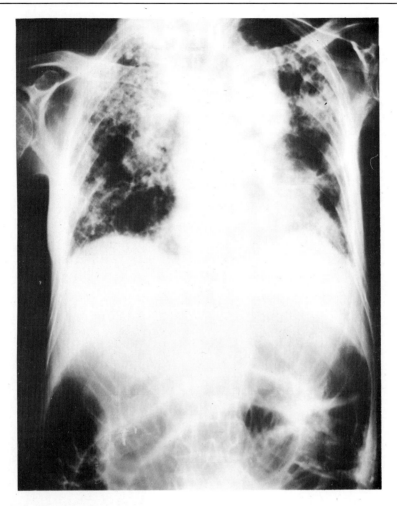

Q2.2 This patient presented with multiple chills and haemoptysis of two days duration.

 (a) Radiological evidence for two disorders is visible — name both disorders.
 (b) What do you think is the most likely diagnosis.

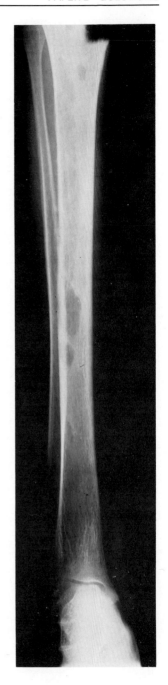

Q2.3 This patient presented with acute abdominal pain. He complained of weakness and change in the shape of his fingers during the previous few months.
   (a) What is the most likely diagnosis?
   (b) Name four possible causes for the abdominal pain.

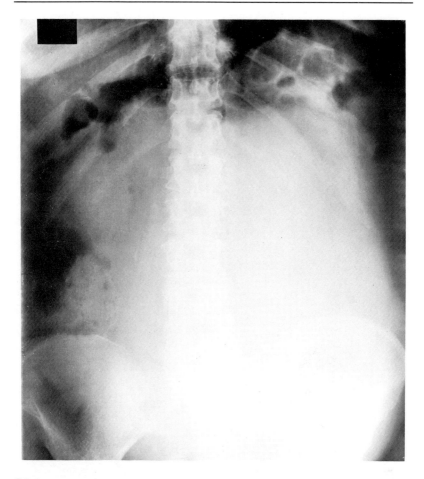

Q2.4

(a) What is the most likely cause of the radiological signs visible in this plain abdominal X-ray?

(b) What investigation would you do to confirm the diagnosis?

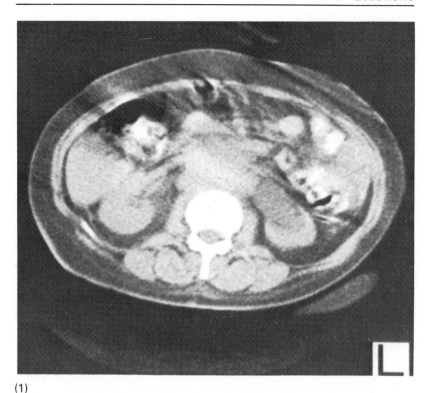
(1)

Q2.5 This 60 year old man presented with weakness in his legs. The CT body scan is a cut through L1.

(a) Name two abnormal radiological signs on the CT scan.

(b) Name the two diseases which may account for both radiological appearances.

ANSWERS PAGE (80)

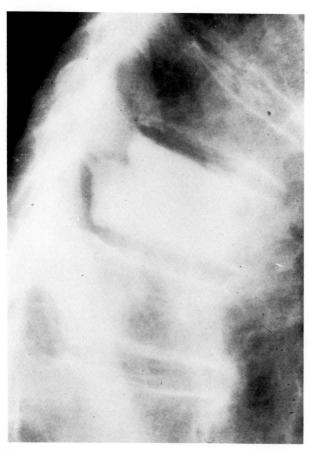

(2)

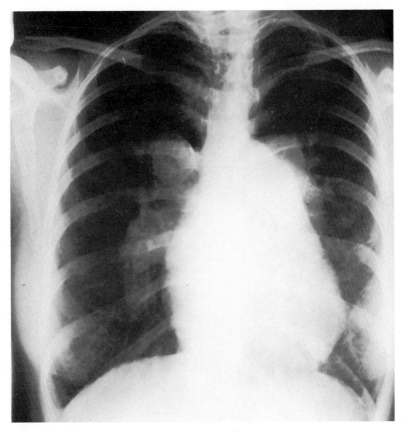

Q2.6 This 30 year old patient was cyanosed.
 (a) What is the most likely diagnosis?
 (b) Name 4 symptoms from which this patient is likely to suffer.

ANSWERS PAGE (81)

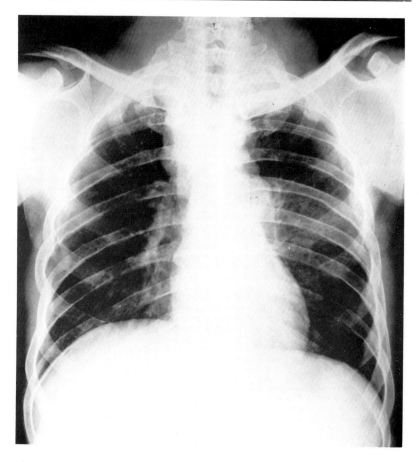

Q2.7 This patient is deaf in one ear.
  (a) Name two abnormal radiological signs.
  (b) What is the most likely diagnosis?

ANSWERS PAGE (81)

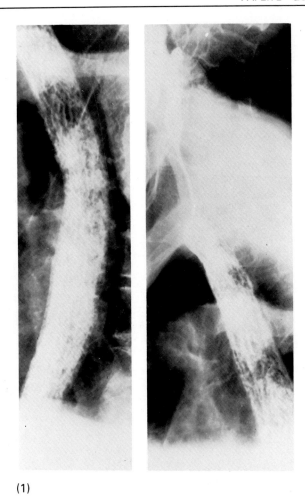

(1)

Q2.8 These three patients are being treated with the same drug.
  (a) What is the drug?
  (b) In each case name the condition which has developed as a result of this drug.

ANSWERS PAGE (83)

PAPER 2 QUESTIONS

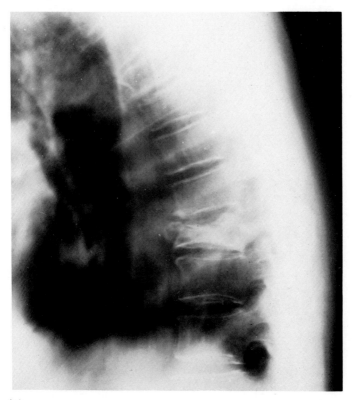

(2)

(3) overleaf

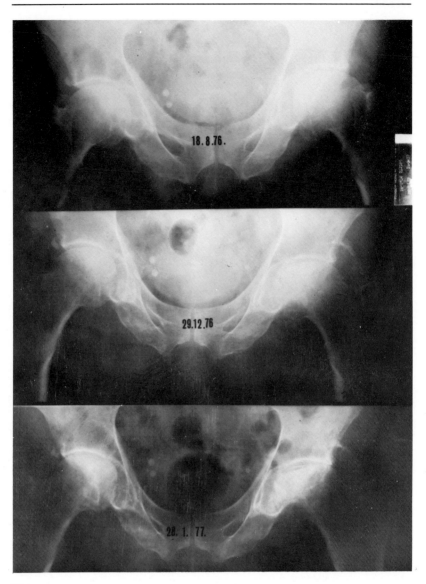

(3)

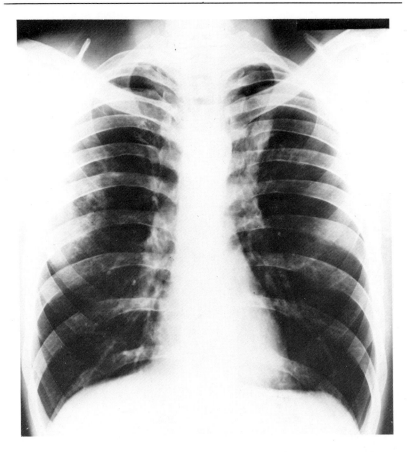

Q2.9 This patient has asthma. During a previous admission similar shadowing was present on chest X-ray but cleared.
   (a) What is the most likely diagnosis?
   (b) Name two tests one would do to confirm the diagnosis.

ANSWERS PAGE (84)

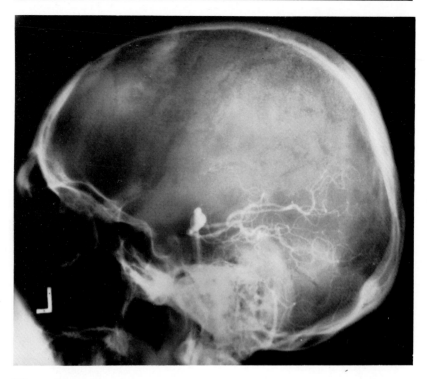

(1)

Q2.10

(a) Name the radiological signs in the vertebral and carotid arteriograms.
(b) Name 2 other disorders frequently associated with this disorder.

ANSWERS PAGE (85)

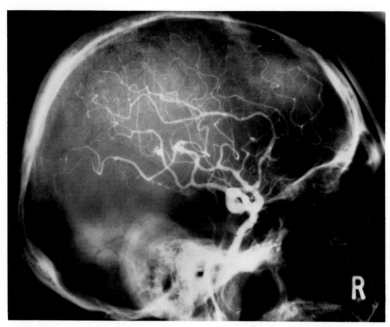

(2)

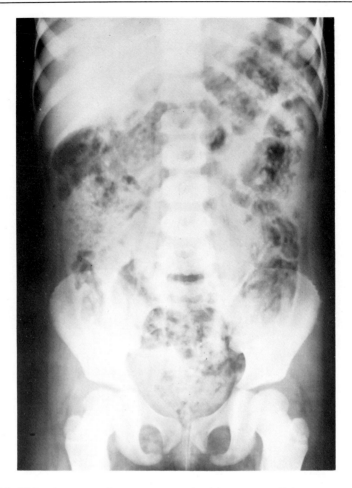

Q2.11 This young patient presented with acute colicky abdominal pain.
  (a) What is the most likely diagnosis?
  (b) How would you confirm the diagnosis?
  (c) How would you treat this patient?

ANSWERS PAGE (85)

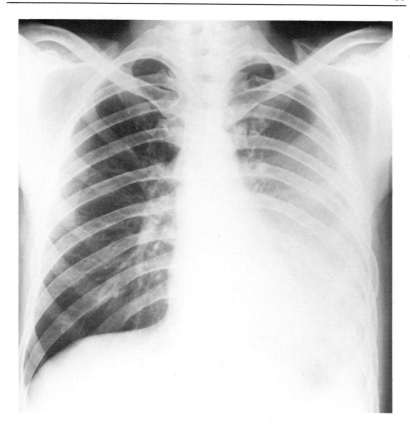

Q2.12 This 30 year old man presented with a 24 hour history of chills and chest pain.

(a) Name the abnormality on his chest X-ray.
(b) What is the most likely diagnosis?

ANSWERS PAGE (86)

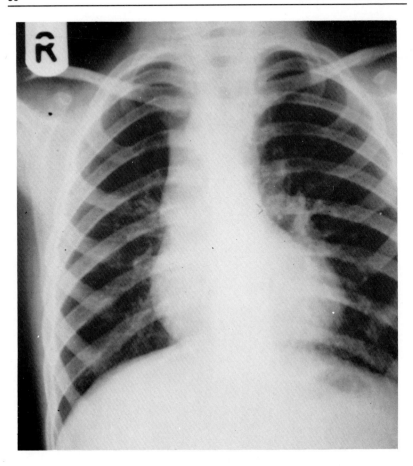

Q2.13 This patient is 15 years old.
  (a) What is the diagnosis?
  (b) Describe the usual auscultatory findings in this condition.

ANSWERS PAGE (89)

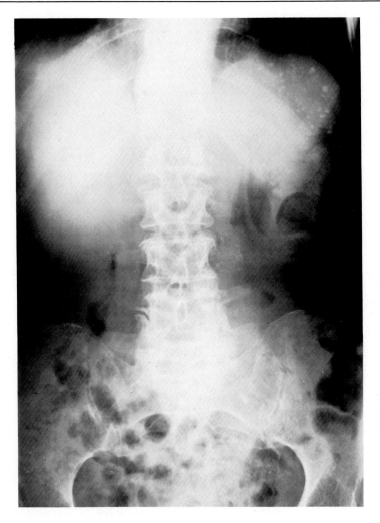

Q2.14 This patient is a chicken farmer in Ohio, USA.
(a) Name the radiological sign.
(b) What is the most likely diagnosis?

ANSWERS PAGE (90)

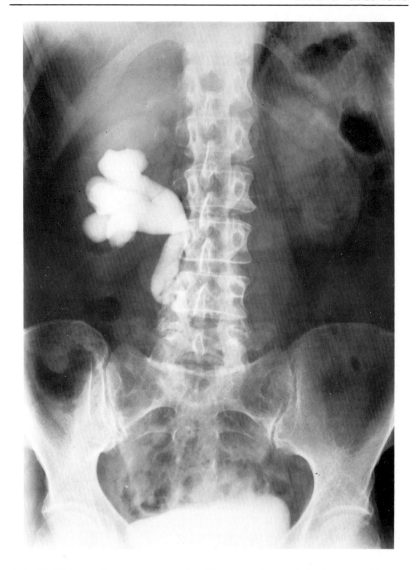

Q.2.15 This patient presented with nocturia, weight loss and low grade fever. This film was taken one hour following injection of contrast material.

(a) What is the most likely diagnosis?
(b) What is the most common symptom of this disorder?

ANSWERS PAGE (90)

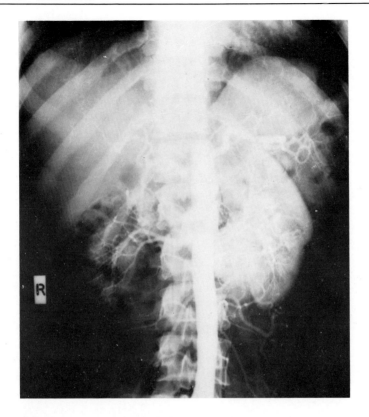

Q2.16 This patient presented with constipation and on examination had postural hypotension.
  (a) What is the most likely diagnosis?
  (b) Name one test you would do to confirm the diagnosis.
  (c) Name two inherited disorders which may be found in association with this disorder.

ANSWERS PAGE (91)

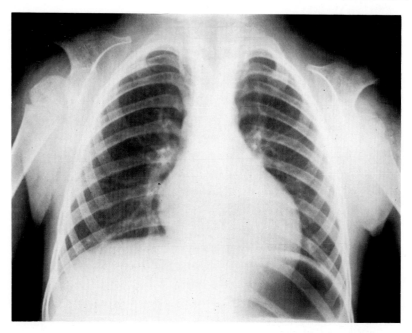
(1)

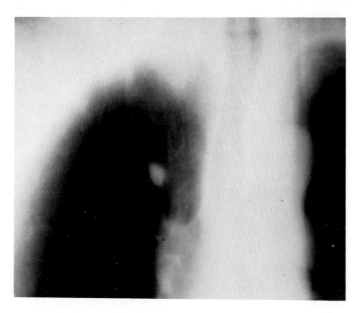
(2)

Q2.17 Spot the abnormality visible on each of the three chest X-rays.

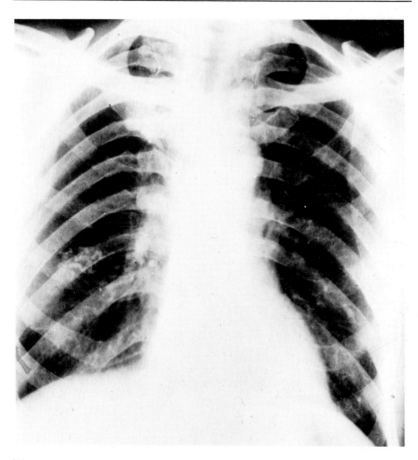

(3)

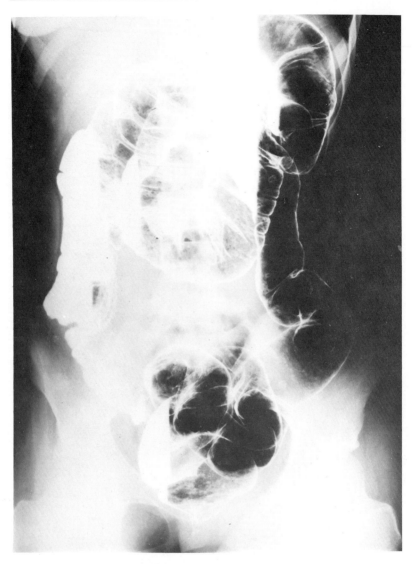

Q2.18 This patient presented with rectal bleeding and was noted to have multiple lipomata.
  (a) What is the diagnosis?
  (b) How would you treat the colonic problem?

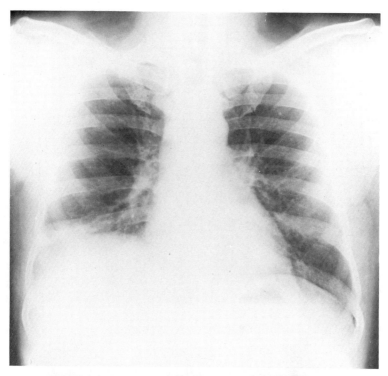
(1)

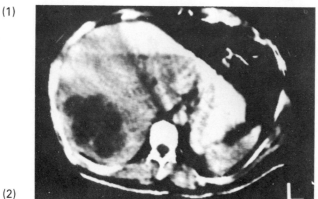
(2)

Q2.19 This patient from Mexico presented with weight loss and fevers. The body CT scan shows a cut through the liver.

(a) What is the most likely diagnosis?

(b) Describe the physical sign which is most likely to be helpful in making a diagnosis in patients with this disorder.

ANSWERS PAGE (94)

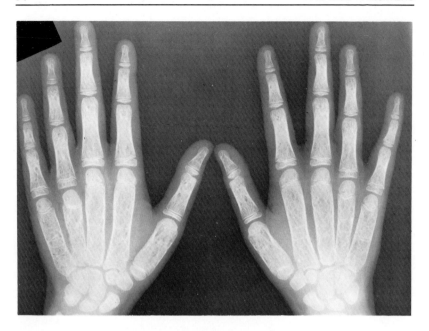

Q2.20 This Cypriot presented with weakness.
(a) What is the most likely diagnosis?
(b) Name the skull X-ray appearance in this condition.
(c) Name five abnormalities commonly observed in the blood film in this condition.

ANSWERS PAGE (95)

PAPER 2

ANSWERS

A2.1 (a) Choledochoduodenal fistula, diverticula of duodenum and jejunum.
   (b) Gallstone ileus

Biliary fistulas are found, often incidentally, in five per cent of patients undergoing gallbladder operations. Seventy per cent of these fistulas go to the duodenum but others go to the colon or stomach, or even abdominal wall. In gallstone ileus the stone usually goes through a cholecystoduodenal fistula rather than a choledochoduodenal fistula and obstructs the small gut at the ileocaecal valve. It is responsible for two per cent of all mechanical small bowel obstructions and usually occurs in the elderly. A history suggesting biliary tract disease is obtained in about half the patients. The diagnosis can sometimes be made from the plain abdominal film which may show small intestinal obstruction, gas in the biliary tree and a calcified ectopic gallstone.

A2.2 (a) Cavitating broncho-pneumonia; intestinal obstruction.
   (b) Staphylococcal pneumonia with paralytic ileus.

Staphylococci which cause bacterial pneumonia reach the lung by haematogenous spread from other organ systems or by aspiration into tracheobronchial tree. Patients with staphylococcal pneumonia are toxic, with multiple chills and high fevers. Chest physical signs in the early stages may not be as impressive as systemic symptoms. The chest X-ray may show progressive bronchopneumonia of airway acquired disease or multiple nodular densities suggesting haematogenous-acquired infections. Abscess formation, pleural effusions and empyema are common. In adults staphylococcal pneumonia may sometimes be recognised when it presents with multiple rounded areas of consolidation which often cavitate. In children staphylococcal pneumonia may result in large tension air cysts, or pneumatocoeles. (Fig. A2.2)

Other causes of cavitating pneumonia include klebsiella, tuberculosis, histoplasmosis and hydatid disease. Severely ill patients with pneumonia commonly get gaseous abdominal distension and sometimes even paralytic ileus. Non-mechanical intestinal obstruction may also occur following myocardial infarction and in the presence of electrolytic disturbances, particularly potassium deletion. However, the commonest causes of paralytic ileus are the result of peritoneal irritation.

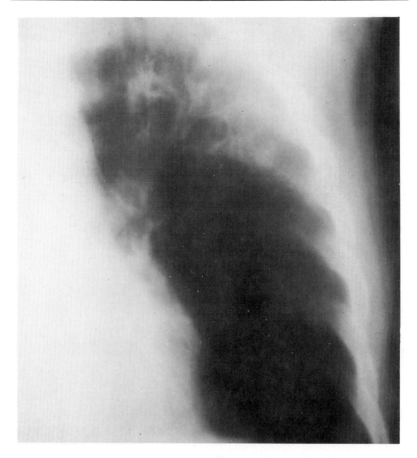

Fig. A2.2 A tomogram of left upper zone showing pneumatocoeles in a patient with Staphylococcal pneumonia.

A2.3 (a) Hyperparathyroidism—osteitis fibrosa cystica.
(b) 1. Renal colic;
2. peptic ulcer;
3. pancreatitis;
4. hypercalcaemia.

Most new cases of hyperparathyroidism show little or no detectable bone disease. However, in some patients the course of the disease is rapid with marked hypercalcaemia, weight loss, debility and bone pain. Radiologically, severe hyperparathyroidism may appear as bone cysts (as in this patient), 'brown tumour' of bone and epulis of bone, in addition to erosive changes in hands, feet and clavicles. Hypercalciuria may lead to nephrolithiasis and passage of calcium oxalate or calcium phosphate stones. Peptic ulcers may result from a potentiating effect of hypercalciuria on acid secretion, or an associated gastrin secreting tumour (Zollinger–Ellison syndrome). Acute pancreatitis lowers serum calcium and thereby masks the presence of hyperparathyroidism, which is associated with both acute and chronic pancreatitis.

A2.4 (a) Ovarian cyst.
(b) Abdominal ultrasound.

A pelvic mass may cause a homogenous soft tissue density shadow arising within the pelvis, extending into the abdomen and displacing the gas filled bowel upwards. Although ascites produces a similar haziness and increase of density, bowel gas shadows are displaced centrally, not upwards, and body wall lines are bowed laterally when ascites is present.
  Ultrasound is a useful investigation in the diagnosis of patients with a pelvic mass because it will demonstrate whether the mass is solid or cystic and whether it is ovarian or uterine. Fig. A2.4 illustrates the clear echo free centre of an ovarian cyst without any internal echos to indicate a solid mass or an ovarian multilocular cyst.

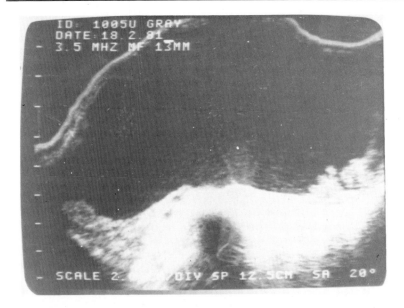

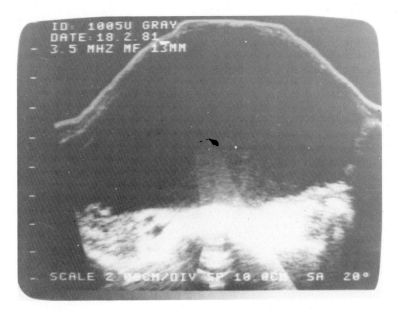

Fig. A2.4 The longitudinal (above) and transverse (below) abdominal echo scans illustrate the echo free centre of an enormous ovarian cyst.

A2.5 (a) 1. Large para-aortic lymphnodes;
2. left hydronephrosis;
(b) 1. Prostatic carcinoma.
2. Lymphoma.

The abdominal CT scan at L1 shows both kidneys: The right is cut through the renal pelvis and the left has low attenuation areas in the collecting system suggestive of hydronephrosis. There is also a solid mass anterior to the vertebra in the para-aortic lymph nodes. The vertebral X-ray shows a dense thoracic vertebra with posterior erosion (suggestive of malignancy) and anterior wedging of the vertebra above. Dense vertebrae are caused by osteoblastic secondaries, especially from prostate and breast, by lymphoma and by Paget's disease.

Prostatic carcinoma can spread by local extension, by the lymphaties to the pelvic and abdominal lymph nodes or by haematogenous dissemination. Bony metastases are the most common form of haematogenous spread and can be osteolytic or osteoblastic. They occur particularly in the pelvis, spine, femur and ribs. Uncommon sites for visceral metastases include lung, liver and adrenal glands. Spinal nerves may be infiltrated by tumour cells or the spinal cord itself may be compressed if the epidural space is invaded.

The complications of retroperitoneal lymphoma are related to obstruction or infiltration of adjacent structures. Obstruction of the ureter(s), particularly the lower third by extrinsic compression is a frequent complication demanding immediate local irradiation. Infiltrative spread anteriorly to the gastrointestinal tract causes lymphatic obstruction of the mesentery or infiltration of the intestinal mucosa. This may manifest itself by nausea, vomiting, abdominal pain, diarrhoea or malabsorption. Posterior spread from retroperitoneal nodes to the spinal cord through intervertebral foramina into the epidural space may result in spinal cord compression.

PAPER 2 ANSWERS

A2.6 (a) Eisenmenger's syndrome due to an atrial septal defect.
  (b) 1. Dyspnoea;
  2. angina pectoris;
  3. syncope on exertion;
  4. haemoptysis.

Eisenmenger's syndrome consists of a reversed shunt where, as a result of any form of communication between the two sides of the heart, blood flow goes from the right to left side of the heart, because pulmonary blood pressure is above systemic levels. The reversed shunt is usually through an atrial septal defect, ventricular septal defect or persistent ductus arteriosus. It is difficult on clinical grounds to diagnose the site of the shunt.

Dyspnoea is probably due to chemoreceptor stimulation by the desaturated arterial blood. Angina and syncope on effort result from the low cardiac output and haemoptysis commonly results from pulmonary infarction.

On the chest X-ray in Q2.6 the main pulmonary trunk and its large branches are very dilated owing to the high pulmonary arterial pressure, but the peripheral lung markings are diminished as a result of high pulmonary vascular resistance. The abnormal cardiac silhouette is the result of a large right atrium and ventricle and small left ventricle and aorta.

A2.7 (a) 1. Superior mediastinal shadow;
  2. rib notching;
  3. upper zone pleural nodules.
  (b) Neurofibromatosis.

Neurofibromatosis is a congenital defect of the ectoderm characterized by dominant inheritance, cutaneous pigmentation and multiple tumours of the spinal and cranial nerves, skin, and internal organs. Neurofibromas appear at any cutaneous site and vary enormously in quality, size and shape. Tumours of the peripheral nerves occur as nodules, usually without interfering with nerve conduction. Cranial tumours are often found, the acoustic and optic nerves being most frequently affected. The signs and symptoms of neurofibromatosis depend upon the localization and rate of growth of the tumours. Tumours lying within the spinal canal or associated arachnoid cysts may produce spinal cord compression (Fig. A2.7).

Neurofibromatosis as a cause of rib notching is very rare, the most frequent cause being coarctation of the aorta, in which dilated intercostal arteries carry blood from collateral vessels to the dorsal aorta, distal to the coarctation. The next most frequent cause is subclavian artery obstruction in which a collateral arterial pathway is developed from the dorsal aorta to the subclavian artery distal to the obstruction. Other rare causes include superior and inferior vena caval obstruction and arteriovenous fistulae.

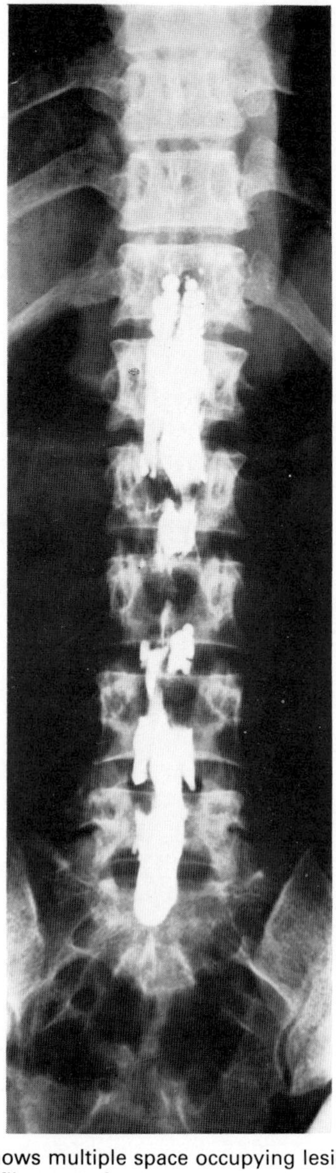

Fig. A2.7 This myelogram shows multiple space occupying lesions in the spinal canal of a patient with neurofibromatosis.

A2.8 (a) **Corticosteroids.**
(b) 1. **Candidiasis of oesophagus;**
2. **osteoporosis with collapse of D8 vertebra;**
3. **aseptic (avascular) necrosis of left hip.**

The reaction to infection is attenuated in patients treated with large doses of corticosteroids or in Cushing's syndrome. Superficial fungal infections, such as tinea vesicolor, occur in about 20 per cent of cases. Candidiasis is also seen frequently in these patients, as it is during therapy with broad spectrum antibiotics, in diabetes mellitus or in patients with immunodeficiency for whatever reason. The mouth is the most frequent site of candidiasis. Within the gastrointestinal tract the oesophagus is the most common site of candidiasis and the only site likely to be symptomatic. Substernal burning, and either pain or a sense of obstruction in swallowing may occur. The radiological appearances are those of small cobblestone or network outline of the barium-filled oesophagus. The lesion is diffuse and the whole of the oesophagus is outlined.

During treatment with steroids insidious atrophy of the skeleton results in mild hypercalciuria and, in time, overt osteoporosis. Severe osteoporosis of the vetebral bodies may result in bulging of intervertebral discs, giving the appearance of codfish vertebrae in lateral X-rays. Compression fractures with anterior wedging of the vertebral bodies result in kyphosis and loss of height. In Cushing's syndrome osteoporosis occurs in over half the patients.

Another frequent skeletal complication of adrenocorticosteroid therapy is avascular (aseptic) osteonecrosis, which most commonly affects the femoral head. Avascular necrosis is due to interference with blood supply to the femoral head which results in degeneration and an arc-like radiolucency on X-ray. Rarefaction of the femoral head occurs, followed by flattening, irregularity and increased density. Finally, the neck of femur becomes short and thick. It heals by recalcification and causes late osteoarthritic changes. This sequence of events is illustrated in Q2.8.

A2.9 (a) Asthmatic pulmonary eosinophilia.
(b) 1. Blood eosinophil count;
2. skin test with aspergillin.

Asthmatic pulmonary eosinophilia is characterized by asthmatic symptoms, recurrent shadows in the chest X-ray and blood eosinophilia. A majority of cases are associated with hypersensitivity to Aspergillus Fumigatus, although other cases may be due to hypersensitivity to Candida albicans. Most patients have developed asthma by the fourth or fifth decade. The attacks may be accompanied by a paroxysmal cough and production of brownish sputum, containing eosinophils and sometimes aspergillus mycelium. There may be accompanying fever, which is sometimes quite high. The attacks may recur at variable intervals and may cease spontaneously. Often the lung shadows are incidental to chronic asthma, but in some patients asthma is present only during an attack and in others it may even be absent. Other organ involvement should suggest the presence of polyarteritis nodosa.

On the chest X-ray shadows are often bilateral, commoner in the upper zones, one shadow clearing to be replaced by others. They may vary from small patches to extensive ones occupying much of the lung field. Usually the shadows are homogenous but may be quite bizarre. It is common to find bronchiectasis in an area of recurrent shadows. In fact the shadow along the left upper mediastinum in Q2.9 occurred in a bronchiectatic area. Occasionally the total white cell count is over 20 000 per mm^3, and the eosinophil count may be as high as 80 per cent. However the count encountered in pulmonary eosinophilia is frequently as high as that which is usual in asthmatics. A high proportion of cases show an immediate positive reaction to the Aspergillus Fumigatus prick test and some show a delayed Type III reaction appearing three to seven hours after the test. Similar Type I and III reactions to Candida albicans can be demonstrated in some cases.

A2.10 (a) 1. Aneurysm at bifurcation of basilar artery in vetebral arteriogram;
2. arterial dilatations at bifurcations in carotid territory.
(b) 1. Polycystic disease of the kidney;
2. coarctation of the aorta.

Ruptured saccular aneurysm is the fourth most frequent cause of stroke, the three more frequent causes being atherosclerosis, embolism and hypertensive intracerebral haemorrhage. The aneurysms are small, thin walled saccules usually located at bifurcations of arteries arising from the circle of Willis. They are presumed to be the result of developmental defects in the media and elastic layer. They vary enormously in size (from 2 mm to 2 cm) and in form. Some are round and connected to the parent artery by a narrow stalk. Others are broad based without a stalk, and some are narrow cylinders.

Saccular aneurysms are rare in childhood and reach their highest incidence between 40 and 65 years. In these patients there is an increased incidence of polycystic renal disease and coarctation of the aorta. Hypertension is more frequently present than in a normal population, but aneurysms may occur in persons with normal blood pressure. Rupture of a saccular aneurysm is characterised by sudden violent headache, collapse, brief unconsciousness and confusion, an absence of prodromal symptoms and few localising neurological signs. In 85 per cent of these patients carotid and vertebral angiography is the only certain means of demonstrating an aneurysm.

A2.11 (a) Ascariasis infection of the gut.
(b) Examine faeces for characteristic ova.
(c) Pyrantel pamoate, mebendazole or piperazine.

Ascariasis is a human infection caused by ascaris lumbricoides and characterized by an early pulmonary phase related to larval migration and a later prolonged intestinal phase. The pulmonary phase may be associated with fever, cough, dyspnoea, wheeze and eosinophilic leukocytosis. Migratory pulmonary infiltrates may be seen during larval passage through the lung. Heavy intestinal infections may cause abdominal pain. Occasionally a bolus of worms may result in volvulus, intussusception or intestinal obstruction, especially in children, because they have smaller intestines and larger worm loads than adults. Moderate malabsorption may also occur.

The diagnosis is usually made by finding characteristic ova in the faeces. This is not difficult as the worm load may be as high as 2000, although it is usually under 50. Each female worm releases 200 000 eggs daily. The fertilized eggs are elliptic with an irregular dense outer shell and a regular translucent inner shell. Occasionally the worms may be seen after a barium meal, either as negative images or after ingesting barium themselves. In Q2.10 the gas filled colon is visible with worms in the lower part of the descending colon.

**A2.12 (a) Left lower lobe consolidation.
(b) Pneumococcal pneumonia.**

The silhouette of diaphragm, heart and great vessels is seen because the air/soft tissue interface produces a difference in radiographic contrast. This silhouette is abolished if air in the lung immediately adjacent to the heart or diaphragm is replaced by fluid or solid tissue. In the left lower lobe two segments (lateral and anterior) are adjacent to the diaphragm. Therefore the left hemi-diaphragm silhouette will be abolished if the left lower lobe does not contain air.

The volume of the lower lobes is best shown by the position of the greater fissures in the lateral projection. Normally the greater fissure on the left side starts at the fourth or fifth dorsal

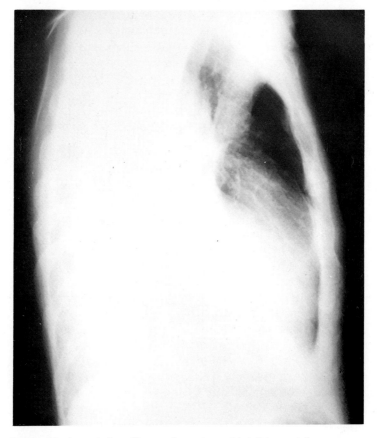

Fig. A2.12a The lateral chest X-ray of a patient with left lower lobe consolidation.

vertebra, passes through the hilum and reaches the diaphragm about 5 cm behind the sternum, as shown in the lateral chest X-ray of the patient in Q2.12 (Fig. A2.12a). With left lower lobe collapse the greater fissure appears to move backwards maintaining the same slope, but passing well behind the hilum (Figs. A2.12b & c). The postero-anterior radiograph may show little evidence of pathology (Fig. A2.12b). When collapse is complete the volume of the shrunken lobe may be so small that it may be hidden behind the heart shadow. Therefore a well penetrated view is important.

The majority of pneumonia's involve the lower lobes. In young adults pneumococcal pneumonia is the most frequent of the bacterial pneumonias. It produces homogenous lobar or sublobar infiltration, often with air bronchograms (Fig. A2.12d). Klebsella pneumonia also tends to be a lobar consolidation but, unlike pneumococcal pneumonia, often cavitates rapidly, may bulge the fissure due to increased volume and often causes emphysema. Viral and mycoplasma pneumonias occasionally cause segmental or lobar consolidation but more frequently cause bilateral diffuse reticulo-nodular infiltrates.

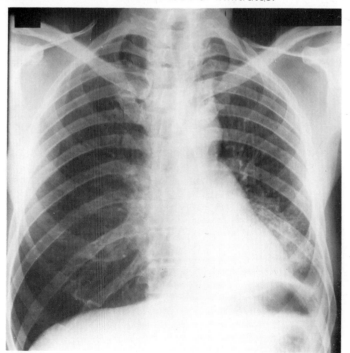

Fig. A2.12b The postero-anterior chest film of a patient with left lower lobe collapse.

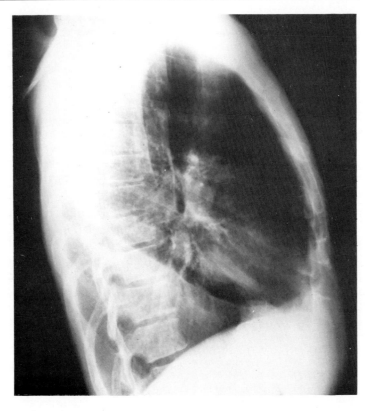

Fig. A2.12c Lateral chest film of a patient with left lower lobe collapse.

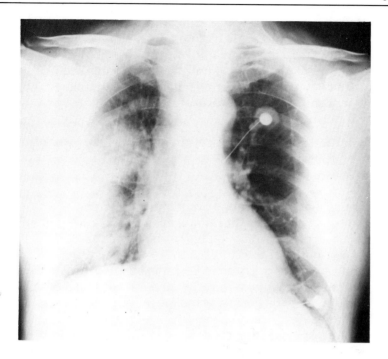

Fig. A2.12d An air bronchogram is visible in the right mid-zone of this patient with bacterial pneumonia.

A2.13 (a) **Tetralogy of Fallot with right-sided aortic arch.**
   (b) **Pulmonary ejection systolic murmur. Single second sound.**

Fallot's tetralogy consists of a high ventricular septal defect, biventricular aorta, pulmonary stenosis and right ventricular hypertrophy. On chest X–ray, the enlarged right ventricle lifts the apex of the heart clear of the diaphragm, sometimes giving the characteristic 'coeur-en-sabot' appearance. If the pulmonary valve rather than the infundibulum is stenosed the pulmonary arterial trunk may be prominent. Usually, however, the pulmonary bay is deep owing to a hypoplastic pulmonary artery. The lung fields are usually oligaemic. When the aorta is right sided the aortic arch may give rise to a knuckle above the right atrium. Sometimes enlarged collateral inter-costal arteries may give rise to unilateral rib notching.

On auscultation a systolic ejection murmur is commonly

audible at the pulmonary area, due to pulmonary outflow obstruction. The murmur is often not long and its maximal intensity occurs in the first half of systole. The second sound is nearly always single because only the aortic component is heard. The low pulmonary arterial pressure makes the pulmonary component inaudible.

A2.14 (a) **Calcification in the spleen (and also along the left border of L3.**
(b) **Histoplasmosis.**

Multiple splenic calcifications occur in histoplasmosis, tuberculosis, brucellosis, haemangiomas and following infarcts. The calcification along the left border of L3 was caused by a colloid carcinoma of the stomach.

Histoplasmosis is a systemic fungal infection of respiratory origin. It spreads by pulmonary lymphatics and blood vessels to form epitheloid or histiocytic granulomas with tubercle-like nodules, caseation necrosis and calcification in lung, lymph nodes, liver, spleen, adrenals and other organs. Man and animals are infected by inhalation of the fungus in dust. Soil from chicken houses or from areas contaminated by bat or bird dung is very rich in organisms. The disease may be asymptomatic; have a primary, acute, predominantly respiratory form; or a progressive disseminated form. The progressive form may be fatal and is characterized by fever, enlargement of liver and spleen, generalized lymphadenopathy, weight loss, anaemia and leucopenia. Sometimes one particular feature may predominate and the patient may present with endocarditis, pericarditis, meningitis, adrenal insufficiency or multiple ulcerations of any part of the gut. Chronic pulmonary disease may also occur which resembles tuberculosis.

A2.15 (a) **Retroperitoneal fibrosis.**
(b) **Low back pain.**

The manifestations of retroperitoneal fibrosis are variable. Vague pain is the most common symptom, usually located in the lower back and sometimes accompanied by symptoms referable to the gut. The patient is likely to lose weight and have low grade fevers. There may be mild anaemia and a high erythrocyte sedimentation rate. Although the ureter is the most commonly affected organ symptoms referable to it are uncommon, until renal impairment develops. The fibrosing process may surround the inferior vena cava but signs of obstruction of that vessel are uncommon. On intravenous pyelography the diagnosis of retroperitoned fibrosis is most often suggested by displacement of the ureters to the midline and evidence of obstruction to one or both ureters usually at the level of the pelvic brim. In this patient's IVP there is delayed excretion of the left kidney and the right is hydronephrotic with the right ureter displaced medially.

A2.16 (a) **Pheochromocytoma**.
 (b) **Plasma catecholamine level**.
 (c) 1. **Neurofibromatosis;**
 2. **Multiple endocrine adenomatosis type II**.

Pheochromocytoma is a catecholamine producing tumour which arises from cells of the sympathetic nervous system. The majority arise in one or both of the adrenal glands but other areas of predilection include the paravertebral sympathetic ganglia (in particular those at the bifurcation of the aorta which make up the organ of Zuckerkandl) and renal pelvis. In Q2.16 the renal arteriogram shows the vessels of the right kidney, the nephrogram of the left kidney and a lesion in the left renal pelvis.

The most common clinical finding is paroxysmal hypertension, which may be associated with palpitations, tachycardia, a feeling of malaise and apprehension, and excessive sweating. However many patients are persistently hypertensive, whereas others have postural hypotension. This latter observation may be found in up to 70 per cent of patients and reflects impairment of normal autonomic reflexes to standing, due to the high levels of catecholamines. Less common clinical manifestations include polycythaemia, impaired glucose tolerance, cholelithiasis and constipation. Severe constipation is probably the result of the marked inhibitory effect of catecholamines on bowel activity and of enhanced sphincter tone. Additional clinical findings may relate to the inherited disorders which may be found in association with pheochromocytoma. Sipple's syndrome (multiple endocrine adenomatosis type II) consists of pheochromocytoma, medullary carcinoma of the thyroid and hyperparathyroidism.

A2.17 **1. Absent clavicles.**
      2. Azygos vein and fissure.
      3. Guinea worm.

Ossification of the clavicles may be completely absent in cleido-cranial dysostosis. Failure of development of the clavicles accounts for the characteristic ability of affected individuals to bring the shoulders together in front of them. In the skull non-mineralization also occurs (Fig. A2.17a) as it may in the teeth and the pubis. With time small islands of bone enlarge and coalesce within the general membrane of the skull, until mineralization is complete. In some individuals, a large anterior fontanel may persist after the separate islands have fused. The relative softness of the calvarium causes it to sag around the base, producing the 'tam-o'-shanter skull'.

The commonest congenital variation in the separation of the lobes is the presence of an extra lobe related to an abnormal position and course of the azygos vein. Normally the azygos vein, as it runs forward to join the superior vena cava, lies medial to the right upper lobe above the right hilum. In patients with azygos lobe the azygos vein remains extrapleural, lies within the right upper lobe, above and lateral to the hilum, and is invested in a double layer of pleura, the azygos fissure. The fissure may be seen in the postero-anterior film running downwards and inwards from near the apex of the lung towards the hilum, the lower end terminating in a small oval shadow, the cross section of the azygos vein.

Infection with the guinea worm usually presents as a skin ulceration at the site of emergence of the female adult worm. Infection occurs on ingestion of infected water fleas, the infective larvae from the flea penetrating the intestinal wall and maturing in the connective tissue under the skin. The female worm takes a year to mature, measuring up to a metre in length and 2 mm in diameter. When ready to discharge larvae she approaches the skin surface, produces a blister, which develops into an ulcer into which the anterior end of the worm protrudes and ruptures on contact with water. Sometimes the female never reaches the surface and calcifies in the tissue producing a characteristic radiological appearance (Fig. A2.17b). Lesions usually occur on the lower leg but may occur on the genitalia, buttocks or upper limbs. Rarely the adult worm involves serous cavities, the extradural space or joints.

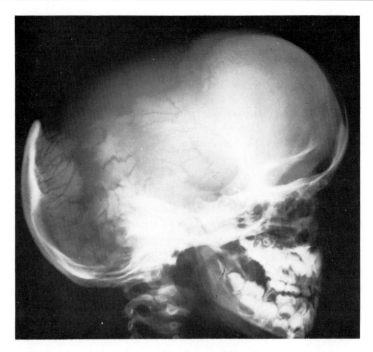

Fig. A2.17a Non-mineralisation of the skull in a patient with cleido-cranial dysostosis.

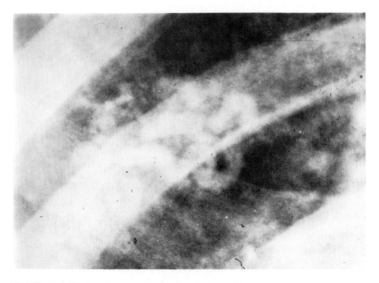

Fig. A2.17b Calcified guinea worm in the chest wall.

A2.18 (a) **Gardner's syndrome.**
(b) **Total colectomy.**

Gardner's syndrome consists of multiple colonic adenomatous polyps and various benign tumours elsewhere, such as lipomas, fibromas, sebaceous cysts, and osteomas. The polyps develop in adulthood and the probability of cancer of the colon occurring is nearly 100 per cent. Therefore total colectomy is recommended. Some patients with Gardner's syndrome have gastric and duodenal polyps which may become malignant. On double contrast barium enema the colonic polyps are seen as multiple small mucosal defects.

Three heritable gastrointestinal polyp syndromes occur, in addition to Gardner's syndrome. (1) Familial colonic polyposis is an autosomal dominant disorder characterized by numerous adenomatous colonic polyps, with a 100 per cent malignant potential. (2) Peutz-Jeghers syndrome consists of multiple hamartomatous polyps of the entire gastrointestinal tract, which rarely become malignant, associated with melanin pigmentation on the skin, lips, and buccal mucosa. (3) Patients with juvenile polyposis have inflammatory polyps in small and large intestine, with no risk of malignancy. These polyps are composed of columnar epithelium, mucous cysts and inflammatory cells.

A2.19 (a) **Hepatic amoebic abscess.**
(b) **Point tenderness in the posterolateral portion of a lower right intercostal space.**

The body CT scan shows a single low attenuation area in the posterior part of the right lobe of liver and the chest X-ray shows an elevated right hemidiaphragm with shadowing above it.

Entamoeba histolytica parasites reach the liver through the portal vein. Usually the abscess is single and is localized in the posterior portion of the right lobe of the liver, because this lobe receives most of the blood draining the right colon, through the effect of streaming in the portal vein blood. This location is responsible for point tenderness in the posterolateral portion of a lower right intercostal space observed in many patients with this disorder, even those without diffuse liver pain. Most abscesses enlarge upward, producing a bulge in the diaphragm, obliteration of the costophrenic angle and, basal atelectasis on the chest X-ray. The right pleural cavity and lung are involved by direct extension from the liver in about a fifth of patients with liver abscess. Many have clinical manifestations related to pneumonia or lung abscess. Perforation into a bronchus occurs with production of large amounts of thick, odourless liquified liver resembling 'chocolate syrup' or 'anchovy paste'.

Isotope liver scan or ultrasound, often revealing a fluid filled cavity with scattered internal echoes, may confirm the presence of a liver abscess. However the diagnosis depends upon the identification of E. histolytica in stools or tissue. In amoebic liver abscess 90 per cent of patients have positive serological tests for the pure antigen. The most important diagnostic procedure is a therapeutic trial of antiamoebic drugs. A dramatic response often occurs within three days.

A2.20 (a) β-Thalassaemia major.
(b) Hair-on-end.
(c) 1. Hypochromia;
2. poikilocytosis;
3. polychromasia;
4. basophilic stippling;
5. target cells.

In β-Thalassaemia major the radiological appearances are striking and are due to increased marrow activity. The long bones are osteoporotic with prominence of trabeculae. The cortices are thin. The medullary cavities are widened, and transverse lines are conspicuous. In the metacarpals and phalanges the expansion of the medullary cavity may produce an almost rectangular contour. In the skull the space between the tables is widened while the outer contour is extremely thin. The trabeculae connecting the inner and outer tables give the appearance of hair standing on end (Fig. A2.20). New growth of the marrow in the upper maxilla reduces the volume (and may even obliterate the maxillary sinuses) and expands the maxillary bone.

In β-Thalassaemia major the anaemia is the result of a decreased production of beta chains. Overall haemoglobin production is decreased and red cell hypochromia results, in which the haemoglobin is limited to a narrow rim at the periphery of the cells. In the blood film marked variation in size and shape of erythrocytes ranging from small microcytes to large pale macrocytes may be seen. Polychromasia, representing regeneration, and basophilic stippling of red cells, indicating haemoglobin denaturation, are also seen. Target cells are present but not to the same extent as in Haemoglobin C disease or sickle cell anaemia.

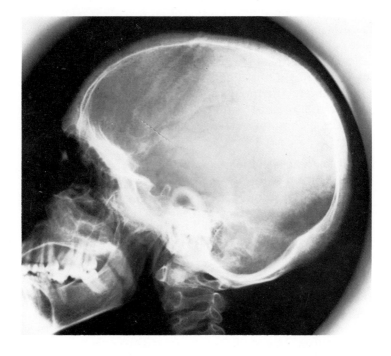

Fig. A2.20 Hair-on-end appearance on the skull X–ray of a patient with Thalassaemia major.

PAPER 3

QUESTIONS

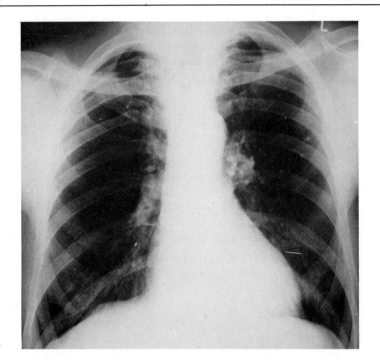
(1)

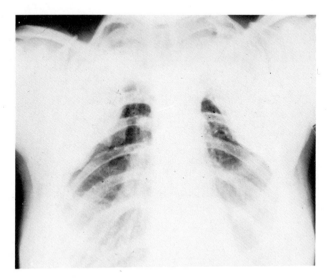
(2)

Q3.1 These patients are suffering from the same disease.
  (a) What is it?
  (b) Name the abnormal radiological sign in each case.

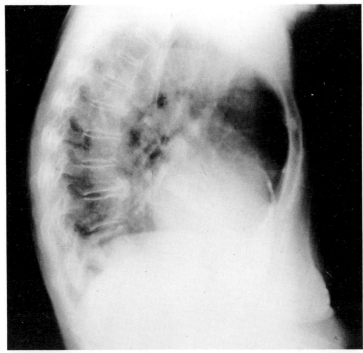

(3)

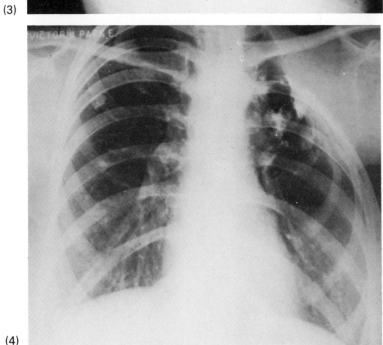

(4)

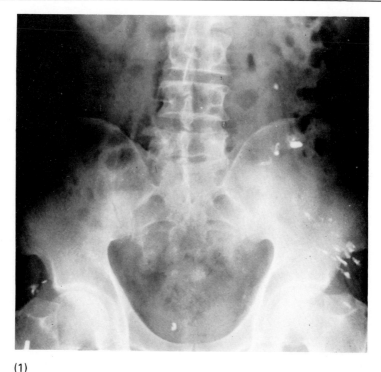

(1)

Q3.2

(a) What is the cause of the radiological abnormality in the abdominal X-ray?

(b) What is the cause of the radiological appearance in the lower limb?

(c) From what disease did this patient suffer?

ANSWERS PAGE (128)

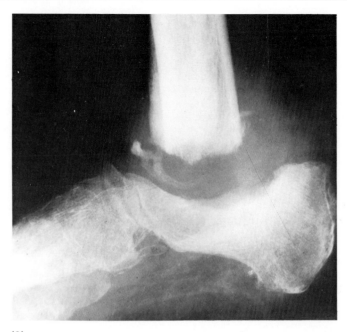

(2)

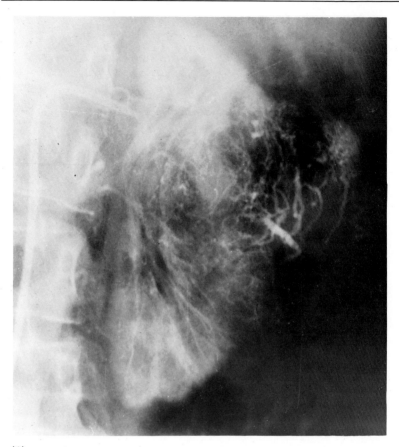

(1)

Q3.3

(a) Name a clinical manifestation which is common to these three patients.

(b) What is the diagnosis in each case?

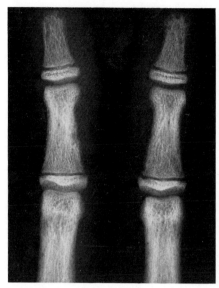

(2)

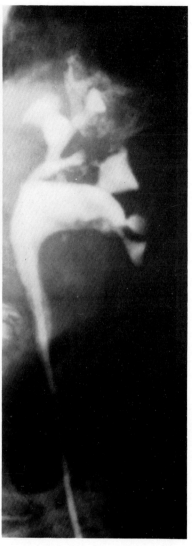

(3)

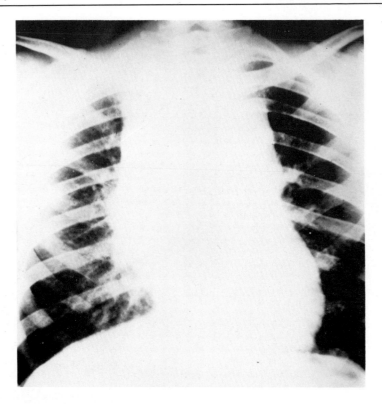

Q3.4 This thirty year old patient presented with a lump in the right supraclavicular area.
   (a) What is the most likely diagnosis?
   (b) Name the crucial cytological feature necessary for the diagnosis.
   (c) What proportion of patients who develop this disorder get alcohol induced pain.

ANSWERS PAGE (129)

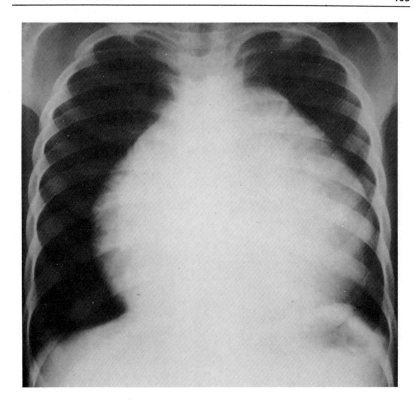

Q3.5

(a) What is the most likely diagnosis?
(b) Name six physical signs which may be present in this patient.

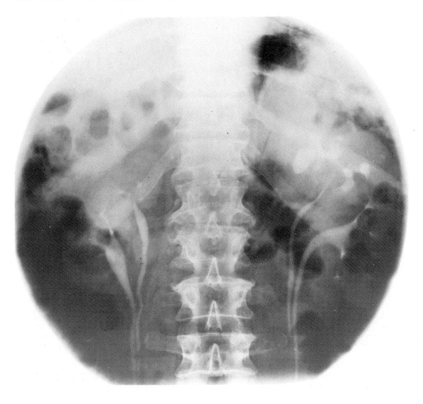

Q3.6

(a) What is the diagnosis?
(b) Name three complications of this disorder from which this patient may suffer.

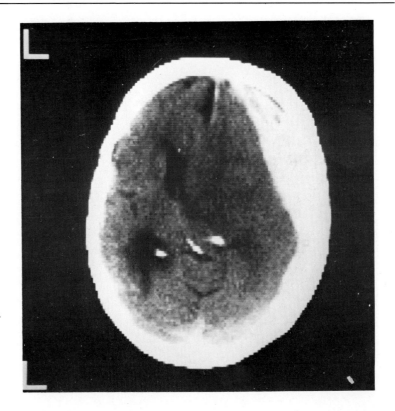

Q3.7 This CT scan was ordered when no acceptable explanation was available for sudden onset of confusion in a 75 year old man.

(a) What is the most likely diagnosis?
(b) What will the carotid arteriogram show?

ANSWERS PAGE (132)

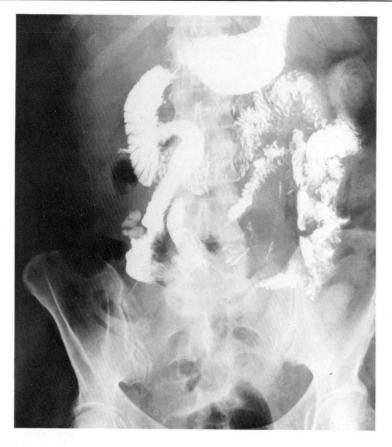

Q3.8 This patient has congestive cardiac failure.

(a) What is the most likely cause of the abnormal radiological signs.

(b) Name three characteristic findings which are likely to be observed in this patient's ascitic fluid.

ANSWERS PAGE (133)

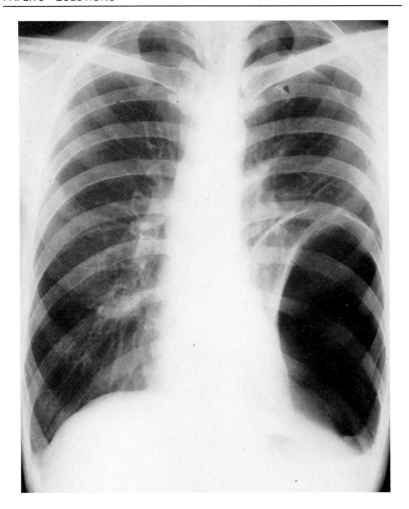

Q3.9 This patient is 50 years of age.
  (a) Name the abnormal radiological signs.
  (b) What is the major predisposing factor to these signs in this patient?

ANSWERS PAGE (133)

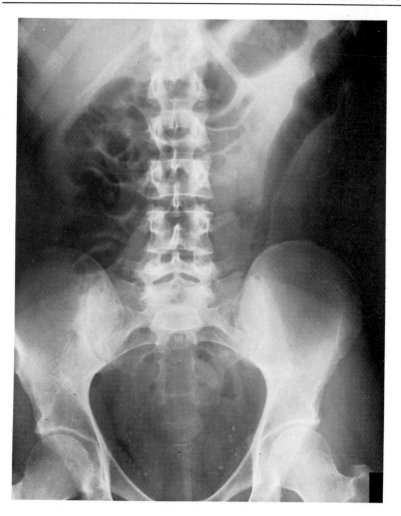

Q3.10 This patient has pruritus.
 (a) What diagnosis can be made from the abdominal X-ray?
 (b) What is the most likely cause of the pruritus?

ANSWERS PAGE (134)

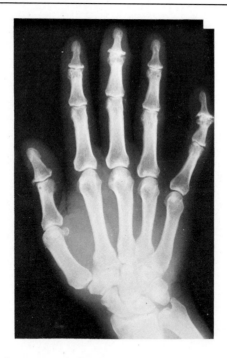

Q3.11 This patient complained of pains in his hands, shoulders and hips.

(a) What is the most likely diagnosis?

(b) Name three physical signs associated with this disorder.

ANSWERS PAGE (135)

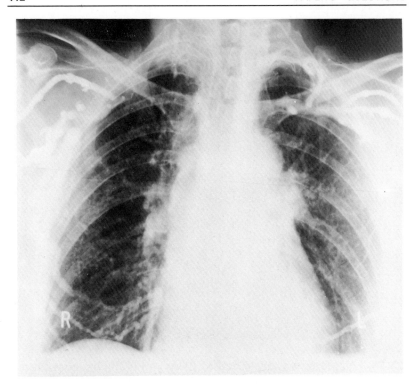

Q3.12 This patient had conjunctival oedema.

(a) Name three physical signs which may be detected in this patient.

(b) Name five causes of this syndrome.

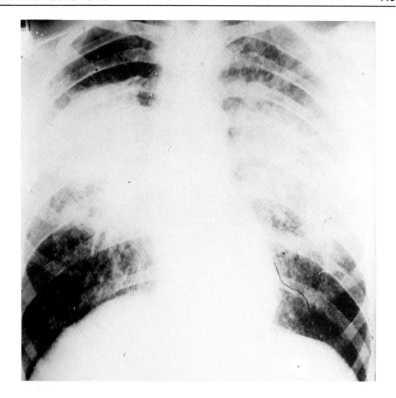

Q3.13 This patient worked in a granite quarry.
  (a) To what dust was he exposed?
  (b) Name the radiological appearance.
  (c) In these patients is there an increased incidence of lung cancer?

ANSWERS PAGE (136)

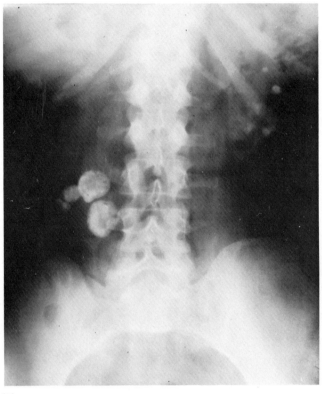

(1)

Q3.14 This patient has rheumatoid arthritis.

(a) What is the most likely cause of this patient's renal disease?

(b) What are the characteristic findings in the urine in this renal disease?

(c) To what tumour are these patients predisposed?

ANSWERS PAGE (137)

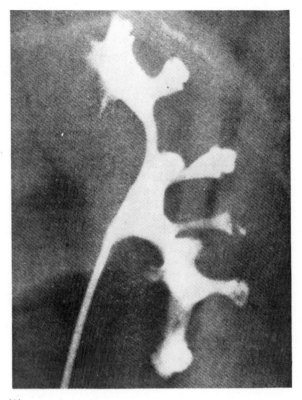

(2)

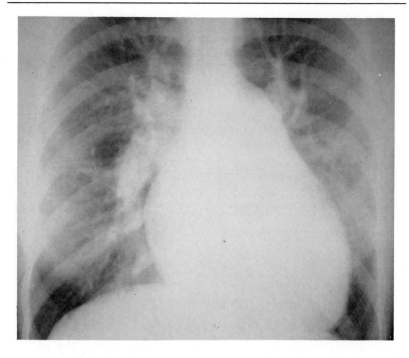

Q3.15 This patient is 35 years old.
 (a) What is the most likely diagnosis?
 (b) Describe the usual auscultatory findings in this condition.

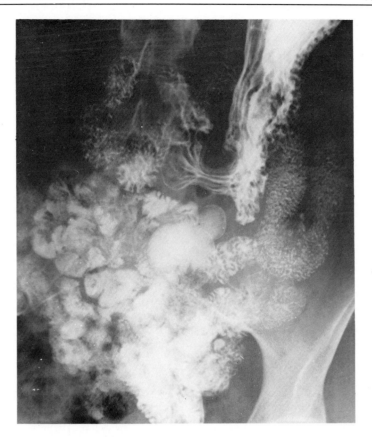

Q3.16 This patient presented with a megaloblastic anaemia.
  (a) Name the abnormal radiological sign.
  (b) Why did this patient develop anaemia?
  (c) Name two tests which would help establish the pathogenesis of the anaemia.

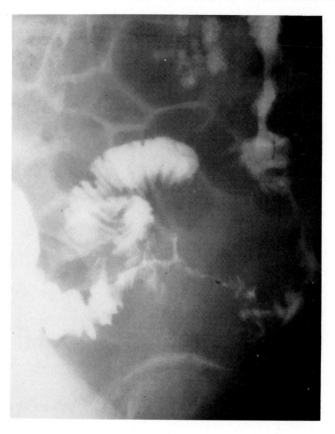

Q3.17 This is a barium enema.
(a) What is the abnormality?
(b) Name the three most common conditions causing this abnormality.

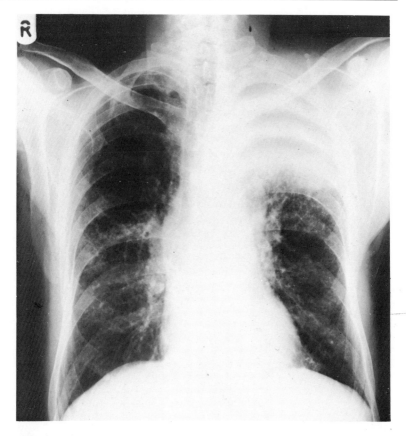

Q3.18

(a) What is the most likely diagnosis?
(b) What ophthalmic signs may be present?
(c) What is the most likely pathological diagnosis?

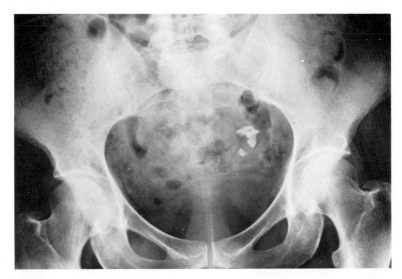

(1)

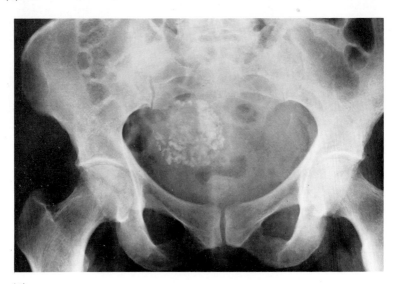

(2)

Q3.19 These three patients had a skeletal survey. Name the abnormal radiological sign in each patient.

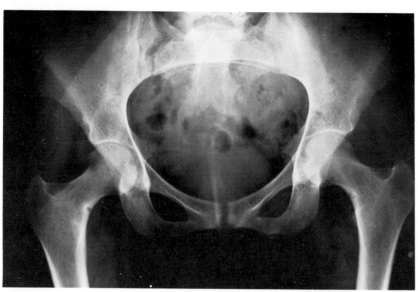

(3)

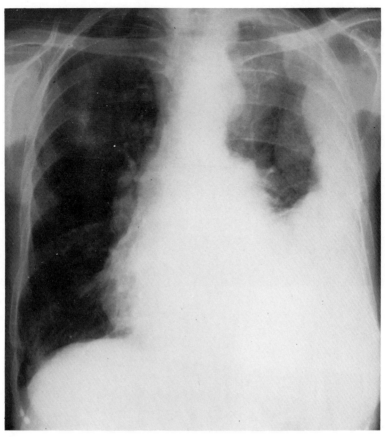

(1)

Q3.20 These two patients were shipbuilders, and have the same disease.
  (a) Name the abnormal radiological signs in each patient.
  (b) What is the diagnosis common to both patients?
  (c) What is the most important piece of advice you can give these patients?

ANSWERS PAGE (141)

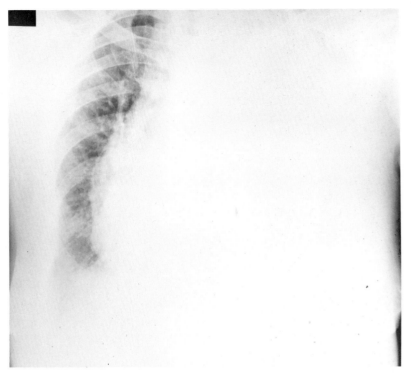

(2)

PAPER 3
ANSWERS

A3.1 (a) Tuberculosis.
  (b) 1. Primary complex;
     2. plombage;
     3. pericardial calcification;
     4. left thoracoplasty.

Tubercle bacilli may gain entrance to the body by several routes, although the most frequent route is by the lung. In the early phase of infection the majority of lesions occur in the lower two thirds of the lungs, where ventilation is best and deposition of droplet nuclei more likely. Organisms from the initial peripheral lesion reach hilar lymph nodes before the progress is inhibited by gradual development of specific immunity over several weeks. This produces epithelioid cell granulomas and caseation necrosis in any site which has bacilli. Infection in the primary site usually heals by a combination of resolution, fibrosis and calcification. The primary complex can

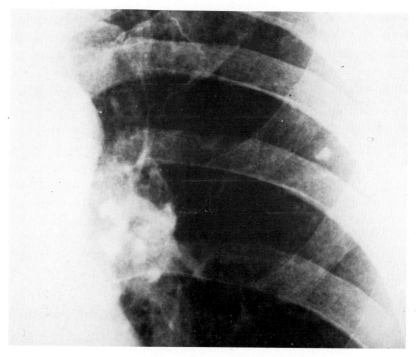

Fig. A3.1a In this patient with tuberculosis a primary complex, consisting of a peripheral Ghon focus and hilar calcification, is visible.

often be seen on chest X–ray and comprises a Ghon focus peripherally and hilar calcification (Fig. A3.1a).

Symptomatic tuberculosis may develop from progression of the primary infection or recrudescence of a dormant lesion. The latter most commonly affects the lung apex. Prior to the advent of effective antibiotic treatment, apical disease was sometimes treated by resection of ribs (thoracoplasty) or by insertion of foreign material (plombage).

Pericardial calcification occurs in nearly half of patients with constrictive pericarditis. It is sometimes difficult to see on the posterio-anterior chest film (Fig. A3.1b) and therefore essential to take a lateral view as calcification is more frequently seen along the diaphragmatic aspect of the ventricles. In the majority of patients with constrictive pericarditis the cause is unknown, although of known causes tuberculosis is the most common. Rare causes include pyogenic infection, following haemopericardium, rheumatoid arthritis, radiotherapy of chest and acute benign pericarditis.

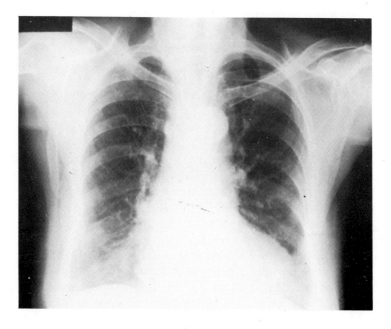

Fig. A3.1b PA chest X–ray showing pericardial calcification.

A3.2 (a) **Bismuth injections.**
     (b) **Neuropathic joint.**
     (c) **Tabes dorsalis.**

Neuropathic joint disease develops when proprioception and/or deep pain sensation is disrupted. Increased trauma occurs during joint movement because of relaxation of the supporting structures of the joint. This leads to cartilage degeneration, recurrent subchrondral bone fractures and proliferation of adjacent bone. The most common cause of neuropathic (Charcot) joints especially when affecting the foot joints is diabetes mellitus. In tabes dorsalis the knees, hips and ankles are usually involved, and in syringomyelia the upper limb joints are damaged. Treatment of the underlying neurological condition seldom influences the progress of the Charcot joint.

In 1943 penicillin was found to be effective in the treatment of syphilis. Subsequently the use of arsenicals, bismuth or mercurials in the management of syphilis became unnecessary. Therefore the radiological sign of multiple linear opacities in the buttocks is now very unusual.

A3.3 (a) **Total haematuria.**
     (b) 1. **Hypernephroma;**
        2. **hyperparathyroidism;**
        3. **urothelial tumour.**

Patient 1 had a selective renal angiogram which demonstrated neovascularity in a tumour, with opacification and pooling of contrast material in the venous sinusoids of the tumour. The finger X-ray of patient 2 shows gross hyperparathyroidism with subperiosteal and terminal tuft erosion. The retrograde ureteric study performed on Patient 3 shows a renal pelvis tumour.

Total haematuria usually indicates that the bleeding occurs throughout the urinary stream and suggests that the bleeding originates from either the kidney or the ureter. Initial bleeding is generally associated with lesions in the urethra distal to the bladder neck and terminal bleeding with the bladder lesions in the area of the trigone. Urinary tract obstructions, calouli and infections account for the bleeding in about 50 per cent of patients with haematuria, whereas tumours alone account for about 20 per cent of all cases. Therefore, it is essential in those patients in whom no other cause is found for the haematuria, to visualize the upper part of the urinary tract by intravenous pyelography and to see the bladder and urethra during cystoscopy. Doubtful lesions in the kidney may be further investigated by nephrotomography, ultrasonography or aortography. Selective renal angiography is reported to have 95 per cent accuracy in diagnosing renal carcinoma. Retrograde visualization of the ureters may be the only method which demonstrates a urothelial tumour, as shown by the third patient who has a renal pelvis tumour, although computerized tomography may also be helpful.

In primary hyperparathyroidism, hypercalcaemia may lead to nephrolithiasis and passage of calcium oxalate or calcium phosphate stones with or without renal colic. Sometimes nephrocalcinosis may develop, which may be complicated by infection.

A3.4 (a) **Hodgkin's disease with hilar and mediastinal lymphodenopathy.**
(b) **Reed Sternberg cell.**
(c) **A sixth.**

The minimal requirement for the pathologic diagnosis of Hodgkin's disease is the presence of characteristic giant cells (Reed-Sternberg) in an appropriate histologic setting. These cells have large inclusion-like nucleoli and double or multiple nuclei of large size. Although Reed-Sternberg cells lack the mitotic index and kinetics of a malignant cell, their aneuploidy, marker chromosomes and invasion of lymphatics and blood vessels suggest a neoplastic origin.

Mediastinal Hodgkin's disease is found in about ten per cent of patients during the course of their illness. A particularly high proportion of right supraclavicular Hodgkin's disease has been observed in patients with mediastinal involvement because the lymphatic drainage of the mediastinum is predominantly to the right supraclavicular lymph nodes. The main complications of mediastinal Hodgkin's are compression of major vessels or bronchi and pulmonary involvement.

Alcohol induced pain has been found in about one sixth of patients with Hodgkin's, although it may occur in other disorders also. The pain is induced by any alcoholic beverage, even if taken in a small amount. It usually occurs within a few minutes and may last up to several hours after alcohol ingestion. The pain is usually localised in the area involved by Hodgkin's disease. It is more frequently observed in patients with mediastinal involvement, leucocytosis, eosinophilia or fibrosis in the tumour tissue than in patients without these characteristics. Alcohol induced pain may precede the clinical or radiographic evidence of Hodgkin's disease. It disappears after effective treatment and may recur at time of relapse.

A3.5 (a) Pericardial effusion.
    (b) 1. Tachycardia,
        2. hypotension;
        3. pulsus paradoxus;
        4. Kussmaul's sign;
        5. stony dull percussion note over heart;
        6. soft heart sounds;
        7. pericardial rub;
        8. Ewart's sign.

In this patient with pericardial effusion the chest X-ray shows a symmetrically enlarged cardiac shadow and clear lung fields.

Normally during inspiration the systolic blood pressure drops by 3–10 mm Hg. When the systolic blood pressure falls by more than 10 mm Hg, the pulse is referred to as pulsus paradoxus, although it is not a reversal, but an exaggeration of the normal response. This occurs when the normal inspiratory increase in venous return to the right side of the heart is prevented by a pericardial effusion, constrictive pericardititis, asthma, or in superior vena caval obstruction. The jugular venous pressure may also show an abnormal pressure response to inspiration in patients with pericardial effusion, in that the jugular venous pressure rises instead of falls (Kussmaul's sign). Stony dull percussion note over the heart is not an early physical sign of pericardial effusion, but, when present, it may be found in the second left interspace. Pericardial friction rub commonly persists, even in a gross effusion. Ewart's sign is dullness and bronchial breathing at the angle of the left scapula.

A3.6 (a) Ureteral duplication with left upper pole hydronephrosis and caculi.
    (b) 1. Hydronephrosis;
        2. calculi;
        3. urinary tract infection.

Minor degrees of duplication are very common. A bifid renal pelvis occurs in about 10 per cent of the population and some degree of ureteric duplication is seen during urography in about 4 per cent of patients. The significance of this malformation lies in the location of the ureteral orifices. One of the orifices is often ectopic and may be obstructed. The intravesical portion dilates to form a cyst or uretocoele, which usually develops in relation to the ureter draining the upper pole of the kidney. The clinical picture is that of urinary tract infection in childhood. The upper pole of the kidney may become hydronephrotic, in which renal calculi may develop, as seen in this patient's control film before IVP (Fig. A3.6). More often this renal segment is non-functioning and dysplastic.

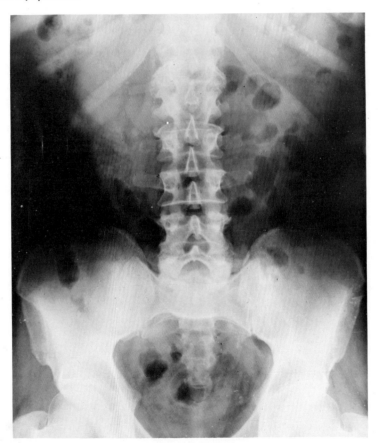

Fig. A3.6 The control abdominal film before intravenous pyelography shows renal calculi in the upper pole of the left kidney of patient in Q3.6 with ureteral duplication.

The ureterocoele can be demonstrated by cystoscopy and retrograde urographic studies. Visualization of the upper urinary tract depends on the degree of residual function. The upper pole may have clubbed calyces, as seen in Q3.6, or may be demonstrated as a thin, opacified rim of tissue, or may not be visualized (due to dysplasia). The lower pole may also be dilated. It is commonly displaced laterally and inferiorly in relation to the ureter, producing the 'drooping lily' sign on IVP.

A3.7 (a) Acute subdural haemotoma.
(b) Separation of middle cerebral artery branches from the skull with contralateral displacement.

The post contrast CT scan shows the surface mass of blood clot and deformity of the ventricular system typical of acute subdural haematoma. The alternative diagnostic technique is arteriography which may show separation of the branches of the middle cerebral artery from the skull with contra lateral displacement. In acute subdural haematoma the traumatic cause is usually clear but may be trivial or even forgotten completely, especially in old people. The main symptoms are headaches, giddiness, slowness in thinking, confusion, and, rarely, fits. Symptoms gradually worsen but may not occur for up to a week following trauma. Focal neurological signs are less prominent than the disturbance of consciousness. Once coma has occurred about half the patients die. The acute subdural haematoma is due to tearing of bridging veins and direct compression of the brain by expanding clot of fresh blood. Chronic subdural haematoma becomes encapsulated by fibrous membranes which grow from the dura. In this encysted state red blood cells and proteins break down, osmotic pressure increases, fluid enters the haematoma which enlarges and compresses the brain.

A3.8 (a) Malignant ascites with peritoneal secondaries.
   (b) 1. Protein > 25 g/l;
   2. high red cell count;
   3. high white cell count with variable cell types;
   4. neoplastic cells.

The follow through barium study shows that the bowel is centrally situated, the loops are separated and there are three filling defects biting into the barium filled bowel, suggestive of malignant ascites. Thorough evaluation, including diagnostic paracentesis, is essential in each patient with ascites, even in the presence of an obvious cause. Thus the patient with congestive cardiac failure may develop ascites from a disseminated carcinoma with peritoneal seeding.

Examination of ascitic fluid should include its gross appearance, protein content, red and white cell count, differential white cell count, gram and acid fast stains, culture, and cytology. Some of these features are sufficiently characteristic to suggest a diagnosis. Blood stained fluid with more than 25 g/l of protein is consistent with tuberculous peritonitis or neoplasia, but the white cells are usually lymphocytes in tuberculosis and are of various types in neoplasia. Cloudy fluid with polymorphonuclear cells and a positive gram stain are characteristic of bacterial peritonitis.

A3.9 (a) 1. Left upper pneumothorax;
   2. large left lower bulla;
   3. emphysematous right lung.
   (b) Chronic obstructive airways disease.

In patients over 40 years, spontaneous pneumothorax is most often due to chronic bronchitis and emphysema. When bullae are present as well, it is usually from a bulla that the air escapes. A bulla has been arbitrarily defined as an emphysematous space greater than 1 cm in diameter. With large bullae the wall of the bulla may be seen radiologically as a fine line, which may merge with the pleura. Bullae rarely displace the heart or trachea but may depress the diaphragm, the upper surface showing a localized small convexity downwards. The characteristic radiographic appearance in pneumothorax is sharply defined lung edge separated from the bony cage by a clear zone devoid of lung markings.

A3.10 (a) **Chronic ulcerative colitis.**
(b) **Sclerosing cholangitis.**

The control abdominal X–ray, prior to a barium enema, demonstrates the featureless colon and absent haustrations of chronic ulcerative colitis, confirmed by the double contrast colonic study (Fig. A3.10).

Extra-colonic manifestations of ulcerative colitis occur in up to 20 per cent of cases, but are rarely the primary reason for operation in ulcerative colitis. Non-articular synovitis is the most common extra-colonic manifestation but sclerosing cholangitis is an important problem. Patients with sclerosing cholangitis

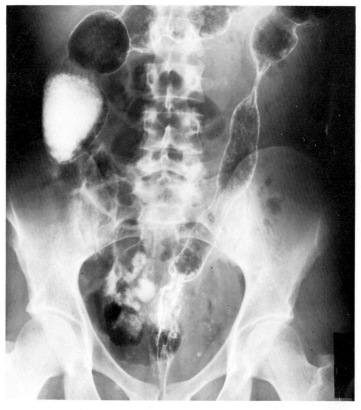

Fig. A3.10 Double contrast barium enema showing the featureless colon and absent haustrations of chronic ulcerative colitis.

usually have longstanding ulcerative colitis which is often relatively quiescent. In some patients abdominal pain, jaundice or pruritus develops, though in most they do not. An elevation of serum alkaline phosphatase without biliary tract abnormalities is the most common hepatobiliary manifestation of ulcerative colitis and does not affect the prognosis adversely. Secondary biliary cirrhosis is very unusual. Chronic active hepatitis may occur in ulcerative colitis, but is also rare.

A3.11 (a) Psoriatic arthritis.
  (b) 1. Nail pitting;
  2. sausage shaped digits from tenosynovitis.
  3. rash.

Distal interphalangeal arthritis distinguishes the erosive arthritis of psoriasis, Reiter's disease and gout from rheumatoid arthritis and other connective diseases which resemble it. Erosive arthritis in gouty patients is usually found in chronic disease and usually associated with tophi (Q1.19). The individual lesions of the joints and soft tissues are similar for psoriatic arthritis and Reiter's syndrome. A distinguishing feature is the distribution of the lesions. Psoriatic arthritis tends to involve both the upper and lower extremities, whereas the arthritis of Reiter's syndrome has a predilection for the lower extremities. In both disorders joint involvement is often asymmetrical and, when bilateral, varied in extent. In the digits severe destruction of the articular surface of the bones comprising the joint may produce a gap between the ends of sharply demarcated stumps of bone. Bone proliferation at the base of the phalanx adjacent to the distal side of the joint is often associated with tapering of the bone on the proximal side of the joint, resulting in a mushroom-like deformity of the joint, sometimes called 'cup and pencil' deformity.

In some patients with psoriasis there is a tendency for severe osteolytic involvement, resorption of bone and telescoping of digits, giving arthritis mutilans. Extensive tenosynovitis results in soft tissue swelling and the digits may resemble sausages. Psoriatic nail involvement occurs in over 80 per cent of patients with arthritis, in contrast to only 30 per cent with uncomplicated psoriasis. However the extent of cutaneous involvement with psoriasis is not necessarily associated with arthritis.

A3.12 (a) 1. Dilated veins of upper thorax and neck with flow of blood towards inferior vena cava;
2. oedema of hands, face and conjunctiva;
3. plethoric face.
(b) 1. Carcinoma of the lung;
2. lymphoma;
3. fibrosing mediastinitis;
4. retrosternal thyroid;
5. aortic aneurysm.

In this patient the bilateral upper limb venogram shows cessation of flow of contrast medium before it reaches the superior vena cava and collateral flow in the intercostal veins below the third rib. The cause of the superior vena obstruction in this patient was never discovered. Obstruction of the superior vena cava is almost always due to malignant disease. Carcinoma of the lung accounts for three quarters of the cases and almost all the remainder are caused by lymphoma. Rarely fibrosing mediastinitis (idiopathic, due to histoplasmosis, or in association with drugs) retrosternal goitre and aortic aneurysms are among the benign causes of this syndrome.

A3.13 (a) Silica.
(b) Progressive massive fibrosis.
(c) No.

Silicosis is a form of pneumoconiosis resulting from inhalation of dust containing silica particles 1–10 μm in diameter. The principal industries associated with silica are mining of gold, tin, copper etc.; quarrying of granite, sandstone and slate; sandblasting; pottery and ceramics; boiler sealing. In fact any industry which involves dust production in the handling of quartz-containing stones or the use of finely divided silica predisposes to silicosis.

In simple silicosis diffuse bilateral miliary or nodular lesions are found, usually more marked in the upper and mid zones. In the complicated form the lesions of massive fibrosis appear as dense shadows (which may cavitate), on a background of nodulation, and may cause distortion of the surrounding tissues.
Differentiation of progressive massive fibrosis from carcinoma and tuberculosis often causes problems. In the past occupations giving rise to a risk of silicosis carried a high mortality from tuberculosis. However silicosis is not associated with an increased incidence of lung cancer.

PAPER 3 ANSWERS 137

A3.14 (a) **Papillary necrosis due to analgesic nephropathy.**
(b) **Raised red and white cell excretion in the absence of bacteriuria.**
(c) **Carcinoma of the renal pelvis.**

Analgesic nephropathy usually results from phenacetin (and sometimes aspirin) abuse. Papillary necrosis may be due to direct toxicity of phenacetin, or a metabolite, or to ischaemia caused by damage to the medullary blood supply. However the picture is frequently complicated by infection, hypertension, stone formation or obstruction. The prevalence of bacteriuria in most series is high, but raised red and white cells in the urine, without bacteriuria, is a well established feature. The unusual site of the carcinoma (renal pelvis) to which patients with analgesic nephropathy are predisposed suggests that some metabolite of phenacetin is carcinogenic.

In papillary necrosis the initial lesion on intravenous pyelogram is a small track of contrast medium entering the calyceal fornix or papillary tip. This track enlarges and erodes further into the papilla and medulla. Eventually the papilla may detach and be lost or lie in situ. The latter may result in a ring shadow of contrast medium around the detached papilla. Calcification, either within the papilla or medulla may also be seen, as in this patient's plain abdominal film (the calcification on the right is due to tuberculous lymph nodes and on the left to papillary necrosis). The retrograde pyelogram illustrates the eroded papillae and a ring shadow in one of the calyces of a patient with analgesic nephropathy.

A3.15 (a) **Atrial septal defect.**
(b) 1. **Loud split first heart sound.**
2. **Pulmonary ejection systolic murmur;**
3. **Fixed splitting of second sound.**

In atrial septal defect, the most striking radiological sign in adults is gross dilatation of the pulmonary artery and its large branches. Fluoroscopic pulsation of these vessels is prominent — hilar dance. The right atrium and right ventricle are large and the aorta and left ventricle are hypoplastic. Following surgical treatment the pulmonary plethora and cardiac size decreases (Fig. A3.15). In children the cardiac silhouette is not as typical

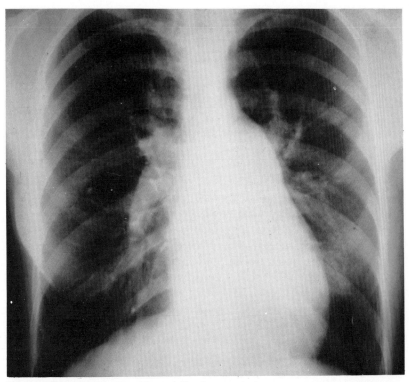

Fig. A3.15 This is the chest X-ray following surgical treatment of the patient with atrial septal defect illustrated in Q3.15. Note that the pulmonary plethora and cardiac size has decreased considerably.

and may be difficult to differentiate from a persistent ductus arteriosus or a ventricular septal defect.

On auscultation the first heart sound is often accentuated and split with a loud tricuspid component. Sometimes a loud atrial murmur precedes it. The second sound is widely split and does not close with expiration — fixed splitting. This may occur because on inspiration the increased venous return goes to the common atrium, both ventricles receive an increased diastolic volume, so that both the aortic and pulmonary components of the second sound are delayed and the interval between them remains wide and unaltered. A pulmonary ejection systolic murmur is nearly always present due to increased pulmonary flow. In one third of patients with ostium secundum defect a mid diastolic murmur may be heard due to increased flow across the tricuspid valve, whereas in nearly all patients with ostium primum defect a mitral diastolic murmur is present.

A3.16 (a) **Jejunal diverticulosis.**
(b) **Malabsorption of vitamin B_{12} due to bacterial overgrowth.**
(c) 1. **Schilling test before and after treatment with tetracycline;**
2. **C^{14} bile acid breath test.**

The proximal small gut is usually sterile, the growth of bacteria being limited by normal peristalsis. Any disorder leading to abnormal stasis of intestinal contents may result in bacterial proliferation and malabsorption, as occurs in jejunal diverticulosis. Both deconjugation of bile salts by bacteria and patchy mucosal lesions cause fat malabsorption. The impaired absorption of vitamin B_{12} is probably due to uptake of B_{12} by micro-organisms.

The Schilling test may be useful in establishing a diagnosis of abnormal bacterial overgrowth in the small gut, because the first stage Schilling test is frequently abnormal and does not return to normal after intrinsic factor, but does so following treatment with tetracyclines. The bile acid breath test utilizing ^{14}C cholylglycine is a reasonably reliable screening test for bacterial overgrowth. The excretion of breath ^{14}C occurs earlier than normal in about two thirds of patients with a positive small gut culture.

A3.17 (a) **Fistula involving large and small bowel.**
(b) 1. **Crohn's disease;**
2. **diverticular disease;**
3. **malignant neoplasm.**

This barium enema shows large gut with haustrations, then a narrow lumen in contact with small gut which has valvulae conniventes. Uncommon causes of fistulas involving the bowel are actinomycosis, amoebiasis, tuberculosis, pelvic inflammatory disease, lymphogranuloma venereum, foreign body, radiation therapy, postoperation and trauma. Radiography may establish the diagnosis by demonstrating one of these diseases, but frequently fails to do so. The major clinical problem is malabsorption when large parts of the absorption surface are excluded from the main stream by large fistulae.

A3.18 (a) **Pancoast's tumour.**
(b) **Horner's syndrome.**
(c) **Squamous cell carcinoma.**

The chest X-ray shows a left upper zone shadow with erosion of the left first and second ribs. Tumours arising in the apex of the lung have been called Pancoast tumours, and are nearly always squamous cell carcinomas. These tumours at an early stage tend to invade surrounding structures, such as bone, pleura, chest wall, brachial plexus and cervical sympathetic chain. Extension to the latter structure may result in Horner's syndrome — enophthalmos, ptosis, miosis and anhidrosis on the side of the lesion. Involvement of surrounding neural structures may produce intractable pain out of proportion to the size of the apical lesion.

A3.19 1. **Teeth in dermoid cyst;**
2. **calcified fibroid;**
3. **looser zone in the neck of both femurs.**

Dermoid tumours consist only of ectodermal structures and may be solid or cystic. Bone and teeth may occur in these tumours which may be discovered accidentally during radiological investigation of other organs. The importance of dermoid cysts lies in their potential for malignant change. Uterine fibroids frequently undergo patchy calcification which is diagnostic on the X-ray film. Pseudofractures (looser zone, Milkman line) are ribbon-like radiolucent zones, perpendicular to free bone surfaces, often bilateral and symmetrical, which give the appearance of incomplete fractures on X-ray. They are found most commonly along the axillary border of the scapula, in the ischial and pubic rami, in the femoral neck, and in the ribs. They tend to occur in bursts and are associated with severe muscle weakness and bone pain.

A3.20 (a) 1. **left pleural tumour.**
2. **Left pleural effusion;**
(b) **Malignant mesothelioma.**
(c) **To apply for industrial injuries benefit.**

Diffuse malignant mesothelioma is nearly always associated with aesbestos exposure. The commonest industries involved are ship building and repairing in men and sackware repairing in women. The induction period after exposure averages 40 years. The commonest presenting complaints are dyspnoea and chest pain. Characteristically the tumour gives rise to a massive pleural effusion with opacification of the lung field and shift of the mediastinum away from the lesion (Q3.20). Sometimes after aspiration of fluid scalloped or ovoid masses may be seen on chest X-ray along the costal margin. Pleural plaques may be present on the contralateral side. Since the prognosis (1 to 2 year downhill course from the onset of symptoms) is not improved by any known therapy the most important advice one should give to patients with suspected meosthelioma is to apply for industrial injuries benefit.

PAPER 4

QUESTIONS

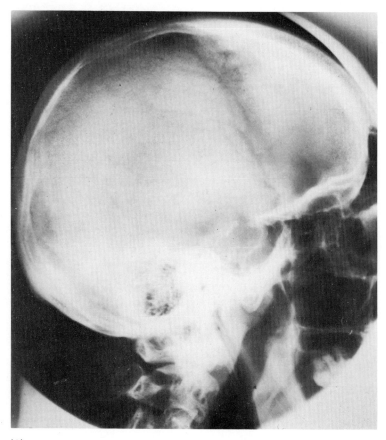

(1)

Q4.1 This patient presented with renal colic.
  (a) Of what disorder are these X-rays characteristic?
  (b) What is the cause of the renal colic?
  (c) What is the syndrome?

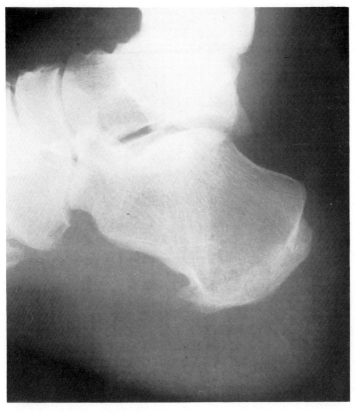

(2)

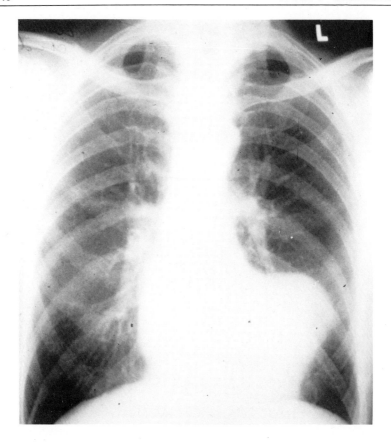

Q4.2

(a) What is the most likely diagnosis?
(b) Name the characteristic physical sign of this disorder.
(c) What is the characteristic electrocardiographic sign of this disorder?

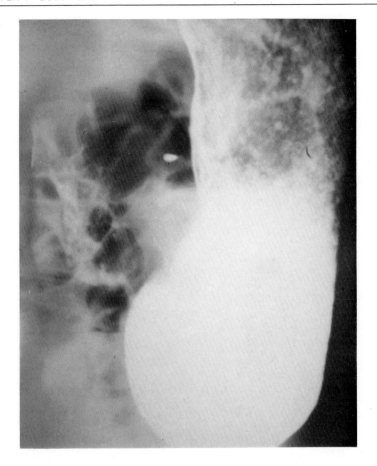

Q4.3 This patient has hypochloraemic alkalosis.
(a) What is the diagnosis?
(b) Name two clinical signs which may be seen in this condition.

ANSWERS PAGE (172)

(1)

Q4.4 These three patients are suffering from the same disorder.
(a) What is it?
(b) What is the most likely cause of the radiological signs observed in each patient?

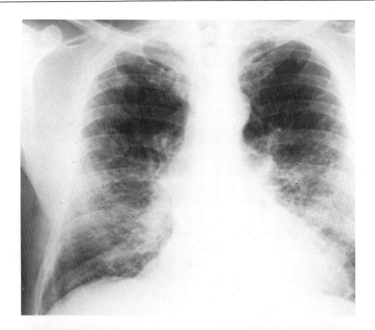

(2)

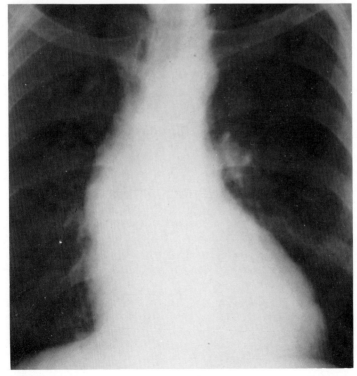

(3)

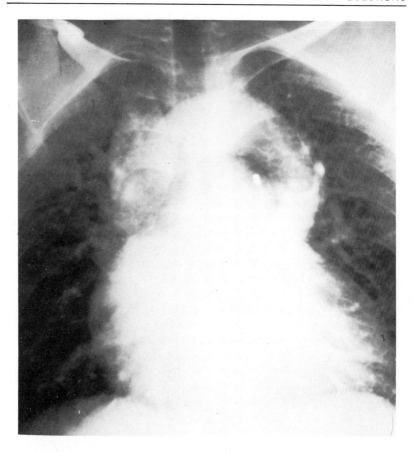

Q4.5 This is the venous phase angiogram of a ten year old child.
  (a) What is the diagnosis?
  (b) What is the oxygen saturation in the right and left ventricles?

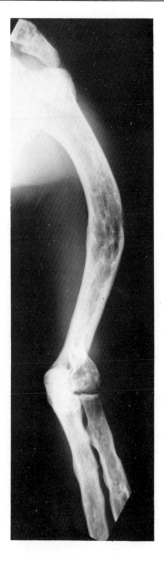

Q4.6 This patient has several areas of dark brown pigmentation with an irregular jagged border.
 (a) What is the diagnosis?
 (b) Name the metabolic bone disease which may be associated with this disorder.

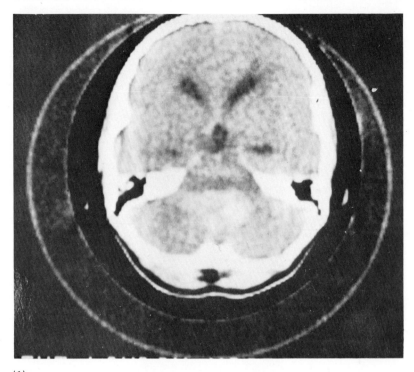

(1)

Q4.7

(a) Name the abnormal radiological signs in both CT scans.
(b) What is the cause of these radiological appearances?

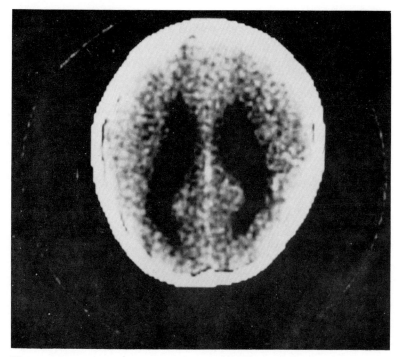

(2)

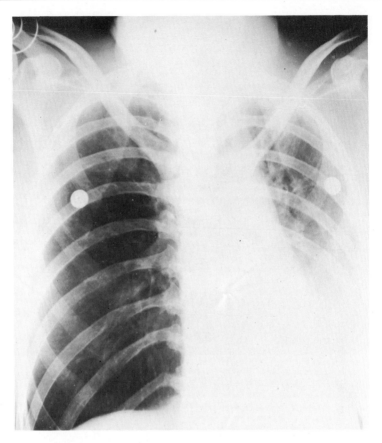

Q4.8 This patient presented with a serum sodium concentration of 120 meq/l. Three months later the patient became dyspnoeic, at which time the above X-ray was taken
(a) What is the abnormality on the chest X-ray?
(b) What is the most likely cause of the hyponatraemia?
(c) What is the most likely underlying pathology predisposing to both these problems?

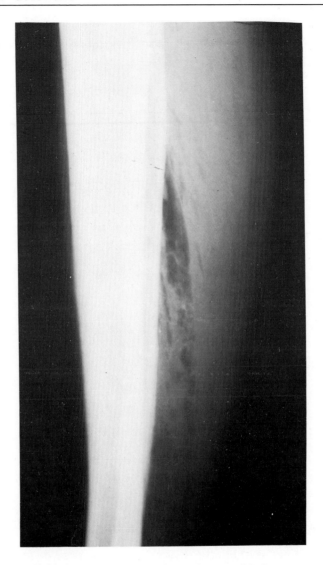

Q4.9 This diabetic patient has a foot ulcer and is in a state of vascular collapse.
  (a) What is the diagnosis?
  (b) Name five therapeutic measures which you would immediately initiate?

ANSWERS PAGE (176)

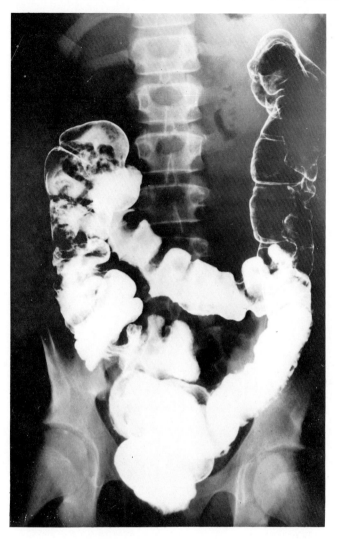

Q4.10 This patient had diarrhoea without urgency or rectal bleeding.

(a) Name the abnormal radiological signs.
(b) What is the most likely diagnosis?

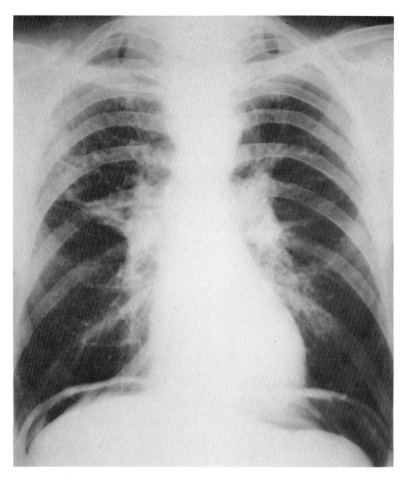

Q4.11 This patient presented with a serum creatinine of 500 μmol/l and sterile pyuria.
 (a) What is the most likely cause of the renal failure?
 (b) Name the radiological signs.

ANSWERS PAGE (177)

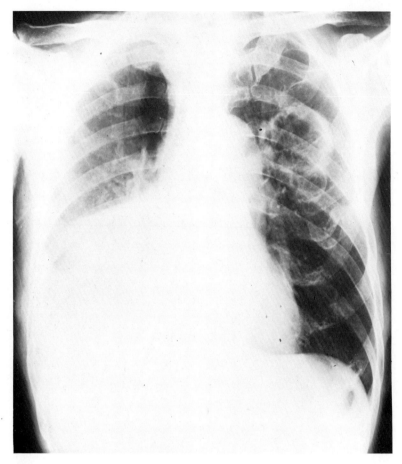

(1)

Q4.12

(a) Name the abnormal radiological signs in both X-rays.
(b) What is the diagnosis?
(c) Name two other causes of the bone abnormality.

ANSWERS PAGE (178)

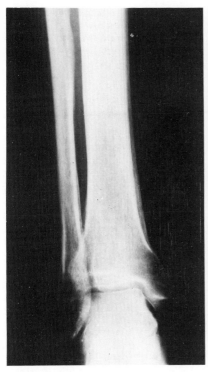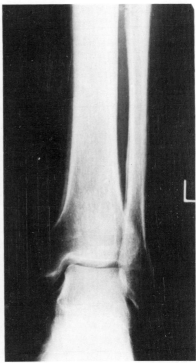

(2)

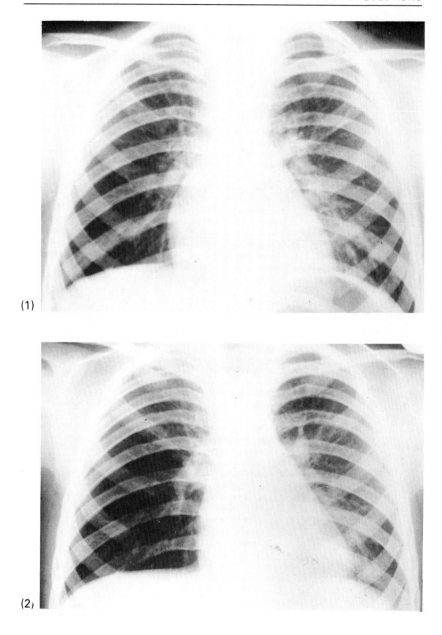

Q4.13 This child complained of retrosternal ache for the previous 10 hours. Inspiratory (top) and expiratory (bottom) chest X-rays were taken.

What is the most likely cause of this patient's pain?

ANSWERS PAGE (181)

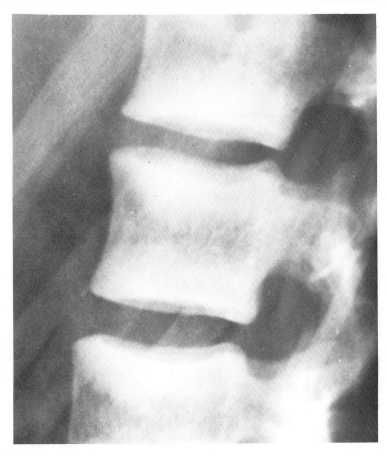

Q4.14

(a) Name the abnormal radiological sign.
(b) What is the most likely cause of this abnormality?
(c) Name three treatments available for this patient's bone disease.

ANSWERS PAGE (181)

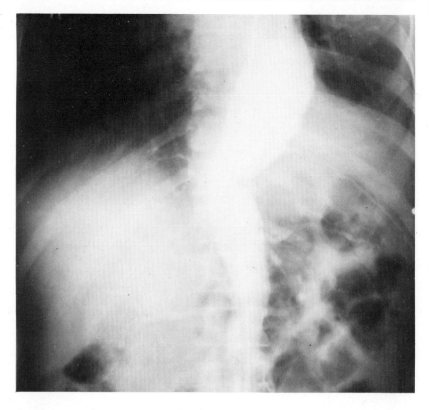

Q4.15

(a) Name the abnormal radiological sign.
(b) What is the most likely cause of this disorder?

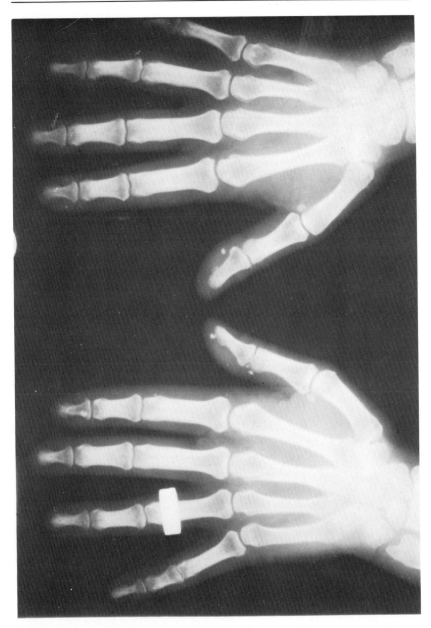

Q4.16 This patient has acrosclerosis.
  (a) Name the radiological sign.
  (b) Name one investigation you would do to confirm your clinical diagnosis.

ANSWERS PAGE (183)

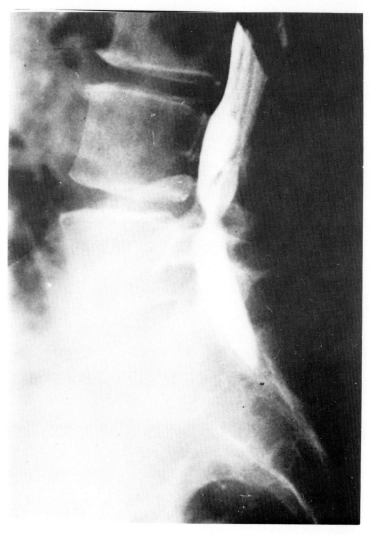

Q4.17

(a) Name the muscles which will be principally affected in this disorder.

(b) Describe the area of sensory loss which may be present in this patient.

ANSWERS PAGE (183)

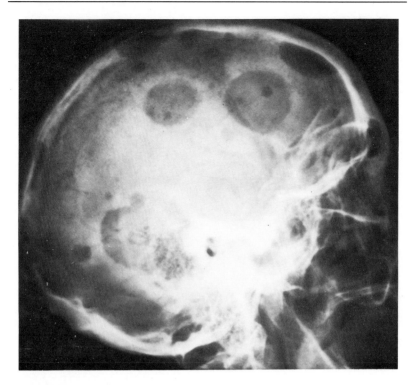

Q4.18

(a) What is the most likely diagnosis?
(b) Name the three most important investigations which may confirm the diagnosis.

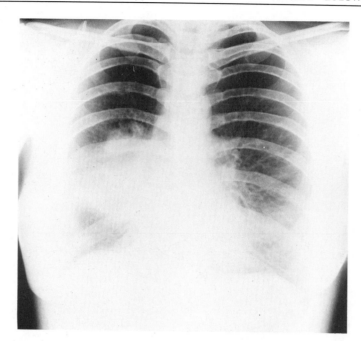

(1)

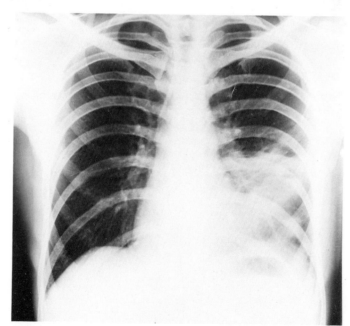

(2)

Q4.19 These three patients have segmental consolidation. Name the segment involved in each case.

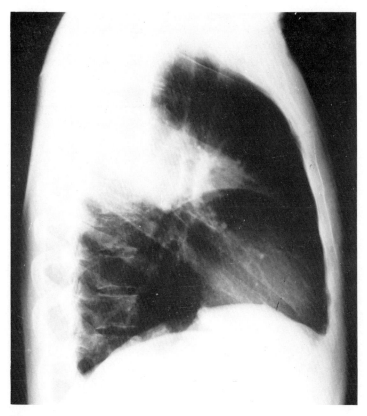

(3)

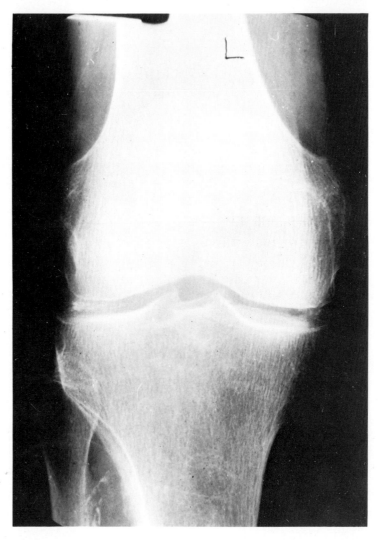

Q4.20 This 50 year old patient with cirrhosis presented with arthritis in his knees and ankles.

(a) What is the abnormal radiological sign?
(b) What is the most likely diagnosis?

ANSWERS PAGE (189)

PAPER 4

ANSWERS

A4.1 (a) Acromegaly.
(b) Hypercalcaemia due to hyperparathyroidism.
(c) Multiple endocrine adenomatosis Type 1.

In this patient with acromegaly the pituitary fossa is greatly enlarged due to the directly expanding effect of an acidophil adenoma of the anterior lobe of the pituitary. In addition to ballooning of the pituitary fossa, erosion and destruction of the anterior and posterior clinoids and of the tuberculum sellae may be seen. A 'double floor' of the sella occurs when the tumour extends downwards unevenly. Plain films of the skull show not only enlargement of the sella turcica in about 90 per cent of patients with acromegaly, but also thickening of the calvarium, enlargement of the paranasal sinuses and widening of the mandibular angle with prognathism. Lateral films of the os calci often show thickening of the heel pad for which the limit of normal is 25 mm. Films of the hands disclose gross increase in the mass of soft tissue, a more characteristic finding then tufting of the distal phalanges (Fig. A4.1). One early sign is an increase in joint space due to cartilage overgrowth. Later degenerative joint changes develop.

Acidophilic or chromophobe adenomas of the pituitary are occasionally associated with functioning islet cell tumours of the pancreas, adenomas of the parathyroids and adrenal cortex, and Zollinger—Ellison syndrome. This disorder, called multiple endocrine adenomatosis Type I, is frequently familial and may present with symptoms of the pituitary tumour, hypoglycaemia, hypercalcaemia (e.g. renal colic) or gastrointestinal haemorrhage. The syndrome called multiple endocrine adenomatosis Type II, comprises medullary carcinoma of the thyroid, pheochromocytoma and parathyroid hyperplasia.

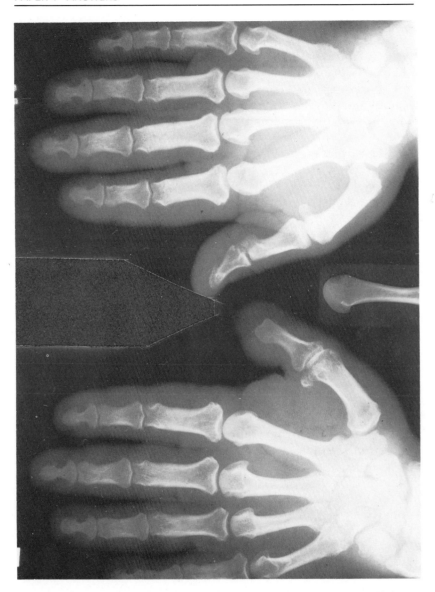

Fig. A4.1 Gross increase in mass of soft tissue and metacarpo-phalangeal joint spaces with tufting of terminal phalanges in a patient with acromegaly.

A4.2 (a) Left ventricular aneurysm.
 (b) Double apical impulse.
 (c) Persistent S-T elevation.

The chest X-ray in this patient with left ventricular aneurysm reveals an abnormal bulge distorting the contour of the heart without overall enlargement. Apical aneurysms are the most common and the most easily detected by physical examination. The most helpful physical sign is a double, diffuse or displaced apical impulse. Congestive cardiac failure occurs in about half the patients with ventricular aneurysms. The electrocardiographic finding of persistent S-T elevation occurs in about two thirds of patients. The definitive diagnostic method is contrast angiography but screening of the heart and radioneuclide ventriculography may also be helpful.

A4.3 (a) **Pyloric stenosis.**
 (b) **Visible peristaltic waves in the epigastrium going from left to right.**
 Succussion splash on ballotement of the left upper quadrant.

This patient's barium meal illustrates the large stomach with long narrowed pyloric canal of pyloric stenosis. Stenosis at the outlet of the stomach is known as pyloric stenosis. It occurs in 2 to 7 per cent of patients with duodenal ulcers and is caused by hypertrophy of the pyloric muscle.
 Hypertrophy has also been attributed to associated antral gastritis or neoplasm and to unrecognised infantile hypertrophic pyloric stenosis. Patients may present with vomiting of undigested food eaten the previous day. The best treatment is probably a limited gastric resection and vagotomy. The gastric resection gives better symptomatic relief than pyloromyotomy and allows an exact histological diagnosis.
 Signs of gastric outlet obstruction may also occur with gastric antrum carcinoma, duodenal bulb peptic ulceration or, rarely, duodenal loop carcinoma.

A4.4 (a) **Rheumatoid arthritis**
 (b) 1. **Amyloid;**
     2. **pulmonary fibrosis;**
     3. **aortic incompetence.**

Amyloidosis occurs in chronic rheumatoid arthritis and may involve the gastrointestinal tract. It may cause ulceration, haemorrhage, protein loss and diarrhoea. Infiltration of the small bowel with amyloidosis gives rise to grossly thickened folds and dilation, (seen in this patient who also has had a left hip replacement).

In rheumatoid arthritis, pulmonary involvement may take the form of diffuse interstitial fibrosis, single or multiple nodules in the lung parenchyma, pleurisy, or pleural effusion. Impaired pulmonary diffusion may occur in the absence of chest film abnormalities.

Cardiac conduction abnormalities and aortic incompetence are occasionally seen, the latter resulting from rheumatoid granulomas in the aortic valve leaflets. Moderate or severe degrees of incompetence are associated with varying degrees of left ventricular enlargement, with the apex displaced downwards and to the left on the postero-anterior chest X–ray. In patients in whom primary valvular disease is responsible for aortic incompetence, the ascending aorta may be moderately dilated and extend further to the right than the right atrial shadow.

A4.5 (a) **Total anomalous pulmonary venous drainage.**
 (b) **Decreased but identical oxygen saturation in both ventricles.**

In this condition blood from the lungs enters the right atrium by one route or another, the most common being into a persistent left superior vena cava, thence into the innominate vein and superior vena cava. To enable oxygenated blood to reach the systemic side an atrial septal defect exists. Identical oxygen saturation in all four chambers of the heart and also the pulmonary artery and aorta is diagnostic of the disorder. More blood enters the right than the left ventricle, because the right ventricle is less resistant to filling than the left and the tricuspid valve is usually wider than the atrial septal defect. Pulmonary blood flow is consequently increased and systemic blood flow is decreased. When pulmonary venous blood drains into the right atrium via a persistent left superior vena cava, the cardiac shadow is pathognomonic and produces a figure-of-eight or cottage loaf appearance (as in Q4.5). As in atrial septal defect the right atrium, right ventricle and pulmonary artery are dilated and pulmonary plethora is present.

A4.6 (a) Allbright's syndrome.
(b) Hypophosphataemic osteomalacia.

Allbrights's syndrome consists of polyostotic fibrous dysplasia, areas of pigmentation, endocrine dysfunction with precocious puberty in females. On X-ray the bony lesions usually comprise a radiolucent area with a smooth border and focal thinning of the cortex. The ground glass appearance reflects the content of thin clacified trabeculae of fibre bone. The lesions may be multiloculate and they may cause deformities such as coxa vara, bowing of the tibia and a leonine appearance. The cutaneous pigmentation tends to overlie skeletal lesions and to remain on one side of the midline. The pigmentation is sometimes described as Coast of Maine because of its irregular jagged appearance. In addition to sexual precosity in females, other endocrine abnormalities observed in patients with Allbright's syndrome include hyperthyroidism, acromegaly and Cushing's syndrome.

A4.7 (a) 1. High attenuation area in posterior fossa;
2. dilated lateral ventricles.
(b) Posterior fossa space occupying lesion causing obstructive hydrocephalus

Hydrocephalus is an increase in the volume of cerebrospinal fluid (c.s.f.) within the skull, and results from:
(a) an increase in volume of c.s.f. without increase in pressure, in which more c.s.f. than usual is present within the cranial cavity in order to compensate for cerebral atrophy (compensatory hydrocephalus) or (b) an increased volume with increased pressure due to a disturbance of the formation, circulation or absorption of c.s.f.

This latter group may be subdivided into obstructive or communicating hydrocephalus. In the former obstruction to the circulation of c.s.f. (either within the ventricles or aqueduct or at the outlet from the fourth ventricle), prevents free communication between the ventricles and the subarachnoid space, whereas in the latter there is free communication between the ventricles and subarachnoid space. Communicating hydrocephalus is due either to increased formation of c.s.f. (choroid plexus papilloma), impaired absorption (high intracranial venous pressure) or obstruction in c.s.f. circulation in the subarachnoid space (inflammatory reaction following infection or haemorrhage). The condition known as normal pressure hydrocephalus is not fully understood. It is possible that patients with this chronic condition have intermittent increases of c.s.f. pressure, and the pressure in the ventricles is reduced as the system enlarges.

The large majority of neoplasms that cause hydrocephalus occur within the posterior fossa or third ventricle. Excepting metastatic tumours these neoplasms are more common in children.

The signs and symptoms of hydrocephalus vary with the age of the patient and depend upon the rapidity with which the intracranial pressure rises and the ventricular dilation occurs. The enlarged ventricles act as a mass compressing and displacing brain tissue.

A4.8 (a) **Partial collapse of left lung.**
     (b) **Inappropriate secretion of ADH.**
     (c) **Oat cell bronchial tumour.**

The chest X-ray shows upper and lower mediastinal shift to the left, shadowing in the left lung with hypertransradiancy in the right lung. Oat cell carcinoma is an anaplastic lung tumour, which is rapidly fatal owing to early metastases, with rapid invasion of mediastinal and more distant lymph nodes. This tumour is capable of producing polypeptide hormones and consequently different endocrine syndromes.

The syndrome of inappropriate secretion of antidiuretic hormone occurs most commonly in patients with oat cell carcinoma but may also occur with other lung tumours, such as adenocarcinoma and alveolar cell carcinoma, carcinoma of the pancreas and duodenum, and in patients with cerebral metastases, which may cause release of ADH by the hypothalamus. To prove the presence of the syndrome it is necessary to demonstrate urine hypertonic with respect to plasma, normal glomerular filtration rate and normal adrenocortical function. Treatment includes severe fluid restriction to less than 500 ml per day, ledermycin and removal of the tumour, if possible. Oat cell tumours have been treated with radiotherapy and combination chemotherapy, although with little success. Hypertonic saline is useful only as an emergency measure for treatment of coma or convulsions due to profound hyponatraemia.

A4.9 (a) Gas gangrene.
    (b) 1. Treatment of shock;
        2. debridement and excision of all dead tissue;
        3. penicillin;
        4. antitoxin;
        5. control of diabetes.

The major features of gas gangrene are those of rapidly progressive muscle necrosis with relatively little inflammatory reaction, rapidly progressive toxaemia, and shock. This is caused by invasion of muscle by clostridia and subsequent elaboration of exotoxins. The factors which predispose to this invasion are (i) impaired local vascular supply (ii) presence of foreign bodies in the wound (iii) presence of necrotic tissue or haemorrhage and (iv) growth of anaerobic microorganisms in the wound. Although the diagnosis of gas gangrene is a clinical one X–rays may show fine bubbles of gas distributed in and around muscle bundles.

    The most important treatment is prompt, thorough debridement and excision of all devitalized tissue and dead muscle. General supportive measures to prevent or treat vascular collapse are necessary, as is penicillin, given in large doses of 10 to 20 million units per day intravenously. Antitoxin is recommended, especially in the first few hours of the disease, in the doubtful hope of neutralizing any free toxin in the body. If available, hyperbaric oxygen may be useful to minimize tissue loss before surgery. In diabetics who have infections blood sugar control may be very difficult to achieve. Consequently in the patient mentioned in Q4.9 another important therapeutic measure would be insulin treatment with frequent blood sugar measurements.

A4.10 (a) **Ulcers in sigmoid colon, normal splenic flexure, Rosethorn ulcers on inferior wall of proximal transverse colon.**
(b) **Crohn's disease.**

Crohn's disease is a chronic granulomatous disorder which most commonly affects the terminal ileum and colon. Affected bowel is sharply demarcated from normal bowel, several areas of disease being separated hy normal bowel (skip lesions). Microscopically the bowel wall is thickened. Affected loops become matted together. Deep fissured ulcers may extend into the submucosa and muscular layers.

The radiological signs of Crohn's disease include narrowing of the terminal ileum, in which barium is seen as a thin irregular tract for the last four to six inches of the ileum (Kantor's string sign); spasm of the caecum and ascending colon (Stierlin's sign); irregular polypoid mucosa giving a cobblestone appearance when seen en face; deep fissures presenting a spiky outline on the barium filled bowel, which, when seen in profile, are called rose-thorn ulcers; skip lesions of normal and abnormal bowel; and bowel of varying diameter with incomplete haustral loss. In contrast ulcerative colitis usually involves the rectum, the lesions are continuous: the bowel is uniformly contracted in chronic cases with absent haustral pattern; the barium outline is granular; if the small bowel is involved, it is in continuity with the colonic disease, and the ileocaecal valve is dilated, as is the terminal ileum, due to backwash ileitis. Clinical features which may help differentiate Crohn's disease from ulcerative colitis are the relative absence of rectal bleeding, presence of fistula formation and perianal or perirectal abscesses.

A4.11 (a) **Tuberculous renal disease.**
  1. **Air under diaphragms;**
  2. **apical shadowing.**

Renal infection is the most common form of extrapulmonary tuberculosis. In about 5 per cent of cases of pulmonary tuberculosis destructive genitourinary tuberculosis will develop, with a latent period of 8 to 40 years, if the pulmonary lesion is untreated. Tuberculous nephropathy is an insidious process which often goes unrecognized for prolonged periods. The finding of sterile pyuria, particularly in an acid urine, may be a clue to the diagnosis. This can be confirmed by culture, or may be strongly suggested by the typical X-ray appearance on intravenous pyelogram. The former is positive in 80 to 90 per cent of cases when at least three separate early morning urine specimens are obtained, and the latter is abnormal in 60 to 90 per cent of patients with tuberculous renal disease.

Extrapulmonary tuberculosis results from the haematogenous spread of tubercle bacilli from a primary source, usually the lungs. However, in more than a quarter of patients with genitourinary tuberculosis no evidence of active or inactive disease can be detected on chest films. Moreover, while the clinically recognisable renal disease may be unilateral, bilateral lesions can be demonstrated patholgically in both kidneys in over 90 per cent of cases with renal tuberculosis.

A4.12 (a) 1. **Periosteal reaction of hypertrophic osteo arthropathy;**
2. **right lower lobe consolidation;**
3. **cavitating lesion left upper zone.**
(b) **Bronchogenic carcinoma with pneumonia.**
(c) 1. **Chronic pulmonary infection;**
2. **cyanotic heart disease.**

Hypertrophic osteoarthropathy is usually associated with bronchogenic carcinoma but may be associated with chronic pulmonary infections, congenital heart disease and hepatic disease. The chest X-ray in this question shows a large thin walled cavity in the left upper zone and right lower lobe consolidation. The latter diagnosis is suggested by loss of right hemidiaphragm silhouette, with retention of right heart border silhouette. The lateral chest X-ray confirms right lower lobe consolidation (Fig. A4.12a). Right middle lobe consolidation is shown in Figure A4.12b bounded by the lesser fissure, with loss of right heart border silhouette and retention of right hemidiaphragm silhouette.

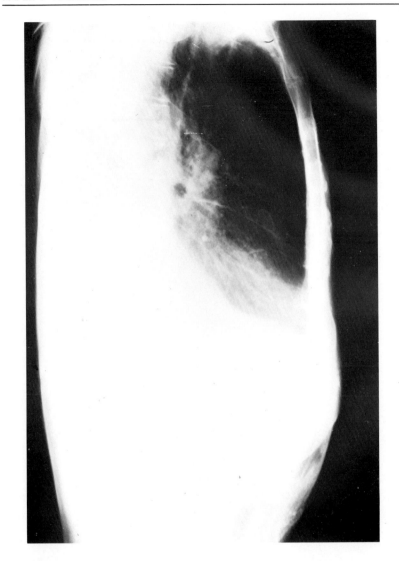

Fig. A4.12a The lateral chest X-ray of patient in Q4.12 shows consolidation in right lower lobe without collapse.

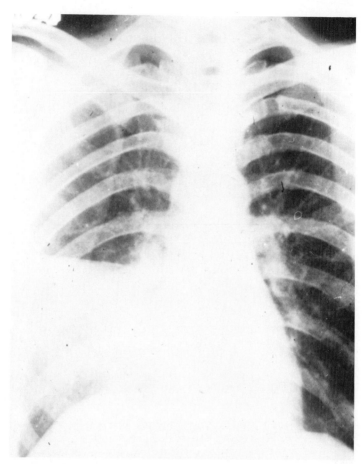

Fig. A4.12b Consolidation of right middle lobe with loss of silhouette of right heart border but retention of the right hemidiaphragm silhouette.

A4.13 Foreign body in right main bronchus.

A foreign body, such as a peanut, in the bronchial tree will cause sharp aching pain, which is increased by coughing. The pain is usually referred to the ipsilateral side of the anterior chest wall and helps to localise the process. A bronchial foreign body may also present with sudden paroxysmal coughing or recurrent chest infections. The radiological signs may be the result of complete bronchial obstruction leading to lung collapse, or incomplete obstruction. In the latter the foreign body may act as a ball valve, so that on inspiration air will be inhaled distal to the foreign body and on expiration the foreign body will obstruct the passage of air out of the lung. Consequently the inspiratory chest X–ray may show subtle signs of unilateral hypertransradiancy and decreased vessel shadowing, but the expiratory film will reveal more obvious hypertransradiancy, with deviation of the mediastinum to the normal contralateral side.

A4.14 (a) Rugger jersey spine.
(b) Renal failure.
(c) 1. Adjustment of serum inorganic phosphorous to normal;
2. vitamin D;
3. parathyroidectomy.

In renal metabolic bone disease there is a predisposition for involvement of the axial rather than the appendicular skeleton. In the spine horizontal bands of increased bone density alternate with relatively less dense bands of bone producing the characteristic 'rugger jersey' spine. It's pathogenesis is uncertain but probably the result of excessive circulating parathyroid hormone. It is usually associated with radiological changes of severe hyperparathyroidism. About half the patients with histological evidence of increased osteoclastic activity have subperiosteal erosions, a radiological sign which is pathognomonic for excess plasma parathyroid hormone. This sign is best observed in the small tubular bones of the hands or feet and the medial margin of the proximal tibia.

If patients with hyperparathyroidism have hyperphosphataemia the first objective is to lower the serum inorganic phosphorus concentration to normal by administration of appropriate amounts of aluminium-magnesium hydroxide gels, which bond intestinal phosphate. Having achieved a normal serum phosphorus 1α -OH-vitamin D_3 or $1,25\text{-}(OH)_2$ vitamin D_3 may be prescribed. If vitamin D does not correct the secondary hyperparathyroidism because of autonomous parathyroid tissue an elective total parathyroidectomy should be considered.

A4.15 (a) Aneurysm of the descending aorta.
(b) Arteriosclerosis.

This arteriogram of the descending aorta illustrates a fusiform aneurysm. A 'true' aortic aneurysm is an abnormal widening of the aorta that involves all three layers of its wall due to destruction of elastic fibres in the media. 'False' aneurysms are disruptions of the inner and medial segments of the wall which permit expansion of the aorta so that the wall of the aneurysm consists only of adventitia and/or perivascular clot. True aneurysms are usually caused by arteriosclerosis. Cystic medial necrosis and syphilis are uncommon causes of this condition. False aneurysms are usually caused by trauma, especially deceleration injuries suffered in car accidents.

The most frequent site of aneurysms is the abdominal aorta, below the renal arteries. The second most frequent site is in the descending aorta, distal to the left subclavian artery. These aneurysms are usually fusiform (i.e. a segment of the artery becomes diffusely dilated and its total circumference is affected). Most patients are asymptomatic when the diagnosis is first suggested by chest X–rays and then confirmed by ultrasound. The risk of elective resection of thoracic aneurysms is about 20 per cent, whereas for abdominal aneurysms it is about 5 per cent. Therefore operative treatment for thoracic aneurysms is indicated only if the aneurysm is symptomatic or expanding rapidly, whereas surgery for abdominal aneurysms is usually undertaken in asymptomatic patients if larger than 6 cm in diameter. Small (4–6 cm) asymptomatic abdominal aneurysms may also have surgical correction if there is no associated, clinically evident, cardiovascular disease.

A4.16 (a) Calcinosis.
(b) Skin biopsy.

The vast majority of patients with scleroderma have acrosclerosis, three quarters of whom have Raynaud's phenomenon or skin tightening as their initial symptoms. The rest usually present with asymmetric polyarthritis. Hand X–rays usually do not show characteristic resorption of distal digital tufts for several years and cutaneous calcinosis may not appear for 7 to 10 years after onset. There are no confirmatory serological tests of scleroderma. Therefore the tentative clinical diagnosis can be supported by skin biopsy, taken from the hand or distal forearm. Early in the disease there may be a mononuclear infiltrate around small arterioles and increased collagen replacing fat in the deeper layers of the dermis. As the cutaneous involvement progresses, there is atrophy of skin appendages with loss of hair follicles and sweat glands, thinning of the epidermis and increased swollen collagen bundles uniformly stained throughout the dermis.

The favourable prognosis of calcinosis in scleroderma is related to the observation that calcinosis does not develop until the disease has been present for 7 to 10 years. The risk of fulminant visceral involvement is associated with diffuse, and often truncal, skin changes, rather than acrosclerosis. Calcinosis affects males and females equally and patients rarely have Raynaud's phenomenon. A variant of scleroderma associated with calcinosis is the CRST syndrome of calcinosis, Raynaud's phenomenon, sclerodactyly and telangiectasis.

A4.17 (a) Dorsiflexors of foot and toes.
(b) Anterolateral aspect of lower leg and dorsum of foot.

The myelogram in Q4.17 demonstrates disc protrusion between L4 and L5. The two most common radicular syndromes of the lower limb involve nerve roots L5 and S1. Differentiation between both syndromes may be made by (a) the distribution of pain, (b) the muscles affected by weakness, (c) the area of sensory loss and (d) the ankle jerk.

(a) In both L5 and S1 lesions pain may occur in the lower back and buttock, but with L5 lesions pain distribution is usually in the lateral thigh and dorsum of foot whereas in S1 lesions it is in the posterior thigh.

(b) The principal muscles affected in L5 lesion are the dorsiflexors of foot and toes, whereas in S1 lesion the calf muscles and plantar flexors are principally affected.

(c) The sensory loss in a L5 lesion is usually in anterolateral aspect of lower leg and dorsum of foot, but in a S1 lesion it is usually along outer edge of foot and sole.

(d) In a L5 lesion the ankle jerk is usually present, although it may be diminished, whereas in a S1 lesion it is frequently absent.

A4.18 (a) Multiple myeloma.
  (b) 1. Tissue biopsy;
  2. bone marrow aspiration/biopsy;
  3. immunoelectrophoresis of blood and urine.

The skull X–ray in Q4.18 demonstrates the characteristic punched-out lesions of multiple myeloma. The basic pathological process in multiple myeloma is the neoplastic proliferation of a single clone of plasma cells. Definite evidence of plasma cell proliferation on tissue biopsy is important but should be accompanied by either a monoclonal globulin spike, lytic bone lesions, or a reduction in normal immunoglobulins. Bone marrow aspiration and biopsy are important in establishing the degree of plasmacytosis and overall marrow cellularity. Since the production of a specific monoclonal immunoglobulin molecule or fragment (M-component) is characteristic of the vast majority of myelomas, the identification and quantification of an M-component is of considerable importance because (i) the amount of M-component is of prognostic importance (ii) serial quantification of M-component is helpful in assessing the response to treatment and (iii) high levels of serum M-component may result in a hyperviscosity syndrome.

Immunoelectrophoresis is a far more specific and sensitive test for urinary light chains than is the classic Bence-Jones heat precipitation method. Urinary M-components are found in three quarters of patients with multiple myeloma and may be the only protein abnormality in a fifth of cases. Therefore a normal serum protein and/or immunoelectrophoresis is not sufficient to exclude a monoclonal gammopathy. However the presence of isolated

Bence-Jones proteinuria is not diagnostic of myeloma and has been found in amyloidosis, macroglobulinaeimia, lymphoma and chronic lymphocytic leukaemia, as well as in occasional individuals without overt disease.

A4.19 1. **Apical segment right lower lobe;**
2. **lingula;**
3. **posterior segment right upper lobe.**

The right lung has 9 segments, three in the upper lobe (apical, anterior and posterior), two in the middle lobe, (lateral and medial) and four in the lower lobe (anterior, lateral, posterior and apical). Consolidation in lung adjacent to the aorta or heart obliterates the silhouette of that part, but consolidation in lung anterior or posterior to it will not do so. Therefore on the postero-anterior chest film shadowing with loss of silhouette along the right side of the aortic knuckle suggests anterior segment upper lobe consolidation; loss of silhouette along the right heart border suggests consolidation of the medial segment middle lobe; and loss of the right hemidiaphragm silhouette occurs in consolidation of anterior and lateral segments lower lobe.

In patient 1 there is shallowing near the pulmonary bay with retention of the cardiac silhouette. A diagnosis of apical segment right lower lobe consolidation was confirmed by the lateral chest X-ray (Fig. A4.19a). In patient 3 consolidation did not obscure the right side of the aortic knuckle suggesting posterior segment night upper lobe consolidation (Fig. A4.19b).

The left lung has 8 segments, 4 in upper lobe (apico-posterior, anterior, superior lingula, inferior lingula) and 4 in the lower lobe (apical, posterior, lateral and anterior). On the postero-anterior chest film shadowing with loss of left aortic knuckle silhouette suggests apico-posterior segment upper lobe consolidation; loss of silhoutte along the left heart border results from lingula consolidation; and loss of the left hemidiaphragm silhouette occurs in anterior and lateral lower lobe segment consolidation.

Patient 2 had shadowing adjacent to the left heart border with loss of the cardiac silhouette. A diagnosis of lingula consolidation (with cavity) was confirmed by the lateral chest X-ray (Fig. A4.19c).

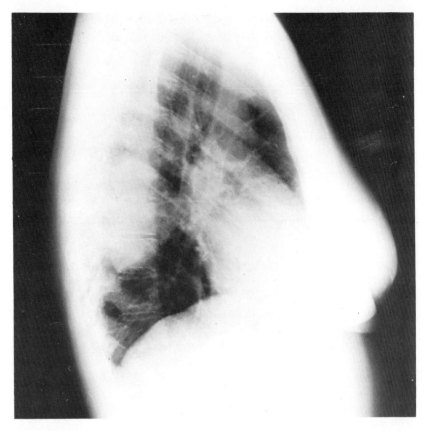

Fig. A4.19a The lateral chest film of patient with consolidation of apical segment right lower lobe.

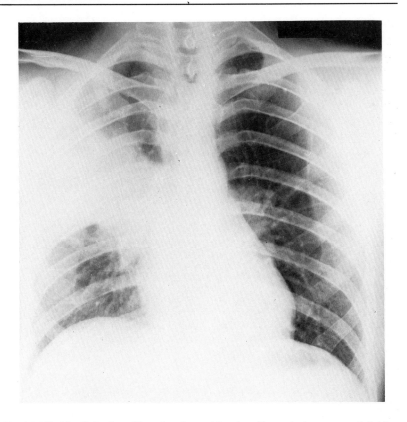

Fig. A4.19b The P.A. chest film of patient with primarily posterior segment right upper lobe consolidation.

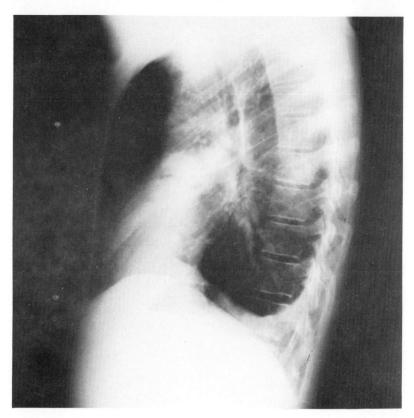

Fig. A4.19c The lateral chest film of a patient with consolidation of lingula in left upper lobe.

A4.20 (a) Chondrocalcinosis.
(b) Haemochromatosis.

Calcification of joint cartilage is called chondrocalcinosis and may be seen in metabolic disorders, such as primary hyperparathyroidism, gout, haemochromatosis. Wilson's disease, acromegaly and ochronosis. However, the commonest cause of chondrocalcinosis in the elderly patient is degenerative joint disease. A familial type and idiopathic condition, associated with calcium pyrophosphate crystals in the affected joints, has been described. These patients suffer from acute recurrent arthritic episodes, giving the disorder the name of pseudogout.

Joint involvement has been reported in as many as half of patients with idiopathic haemochromatosis. The joint disease resembles degenerative disease rather more than rheumatoid arthritis. There is narrowing of the joint space with subchondral erosions and sclerosis. Chondrocalcinosis may occur in about a third of these patients with arthropathy and may account for crystal-induced synovitis, giving acute arthritis.

PAPER 5

QUESTIONS

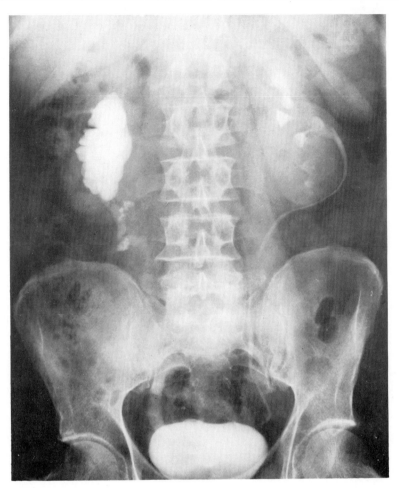

(1)

Q5.1 This patient presented with hypertension.
(a) Name the abnormal radiological signs.
(b) What is the most likely cause of the radiological signs in the right and the left kidneys?
(c) Name two haematological abnormalities which may occur in this patient.

ANSWERS PAGE (218)

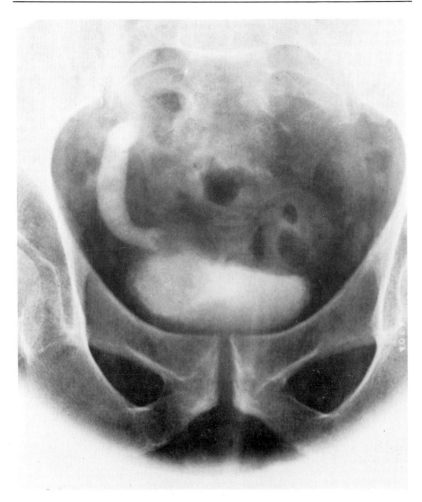

(2)

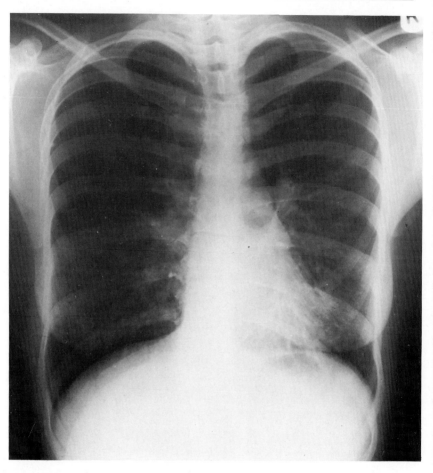

Q5.2 This patient presented with cough productive of sputum for 2 months.
  (a) Name two abnormalities on chest X-ray?
  (b) What is the cause of the cough and sputum?

ANSWERS PAGE (218)

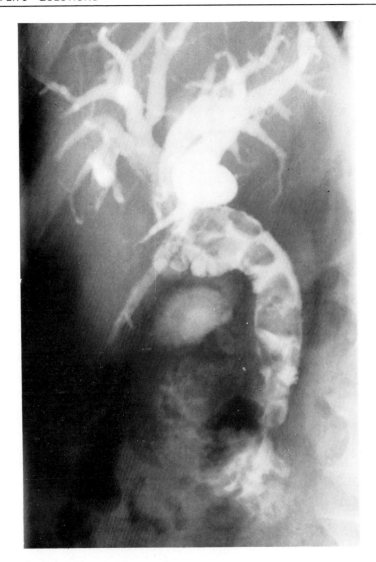

Q5.3 This is a percutaneous transhepatic cholangiogram.
 (a) What is the diagnosis?
 (b) Name five different ways in which this patient may present.

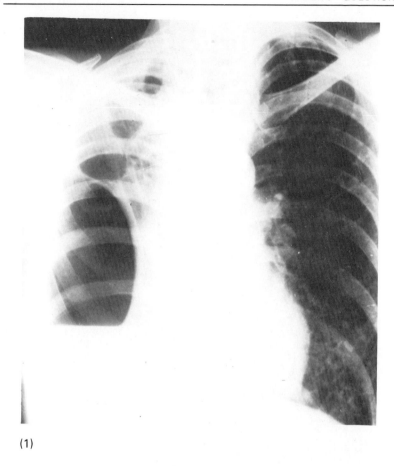

(1)

Q5.4 Name the radiological abnormalities in each chest X-ray.

PAPER 5 QUESTIONS

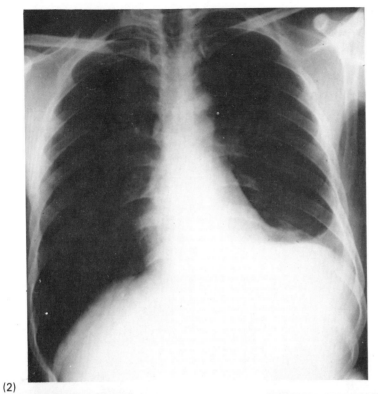
(2)

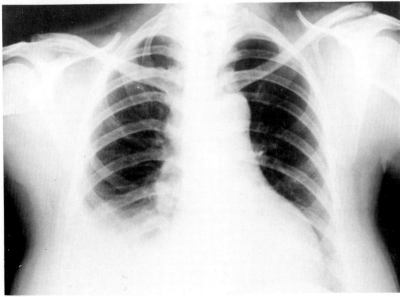
(3)

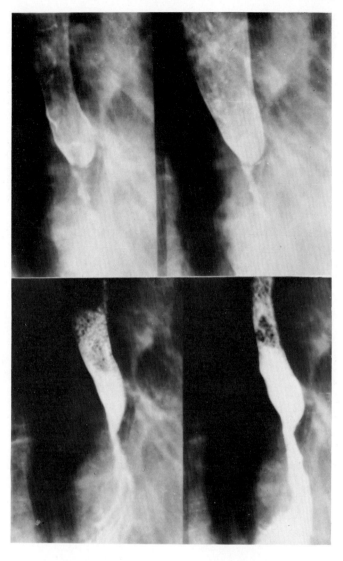

Q5.5 This patient presented with acute dysphagia.
(a) What is the cause of the acute dysphagia?
(b) Name three other ways in which this patient may have presented.

ANSWERS PAGE (222)

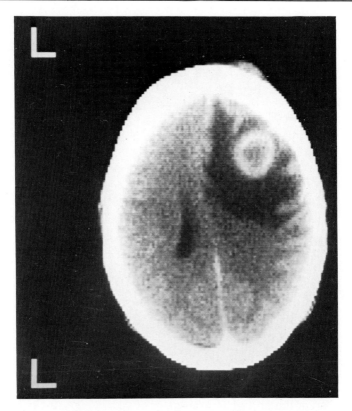

Q5.6 This patient had sinusitis for many years and then presented to casualty with drowsiness and inattention.
 (a) What is the most likely diagnosis?
 (c) How would you treat this patient?

ANSWERS PAGE (222)

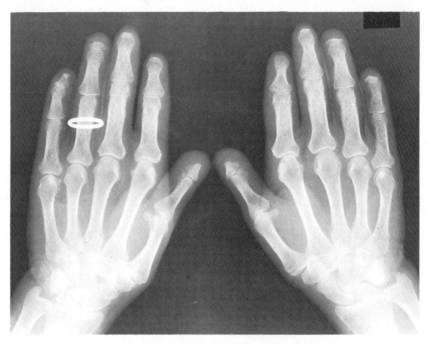

(1)

Q5.7 These 2 X–rays are of the same patient. The chest X–ray was taken at the end of inspiration.
  (a) Name the abnormality on the hand X-ray.
  (b) Name two abnormalities on the chest X–ray.
  (c) What is the diagnosis?

ANSWERS PAGE (223)

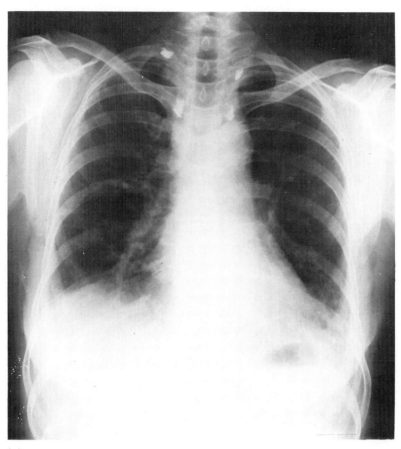

(2)

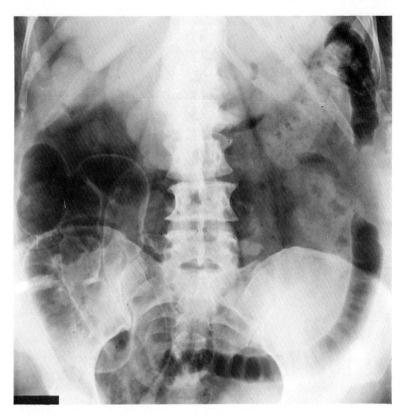

Q5.8 This patient presented with diarrhoea and right upper quadrant pain suggestive of duodenal ulceration. He was admitted two months later with an acute abdomen, despite antacid therapy.

(a) What is the cause of the acute abdomen, judging by the plain supine abdominal film.

(b) Name a disorder which may account for all this patient's problems.

(c) How would you confirm this diagnosis?

ANSWERS PAGE (224)

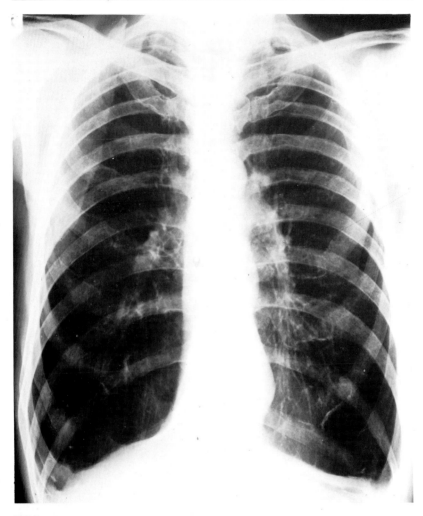

Q5.9

(a) Name this patient's major symptom.

(b) Name the three major predisposing causes to this disorder.

(c) Describe the abnormalities of pulmonary function (in tests which are routinely available) which you are likely to find in this patient.

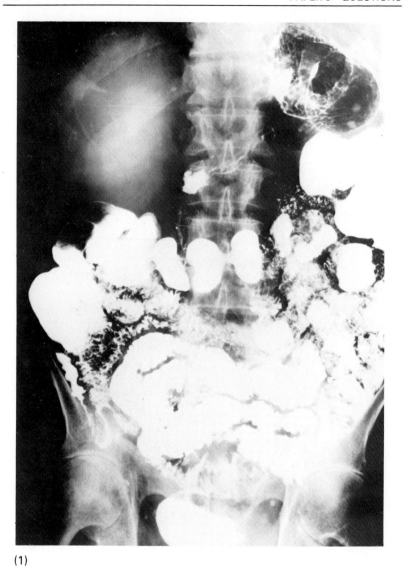

(1)

Q5.10 These three patients have the same disease.
 (a) What is it?
 (b) Name the abnormal radiological sign in each case.
 (c) What is the most likely presentation in patient 1?

ANSWERS PAGE (225)

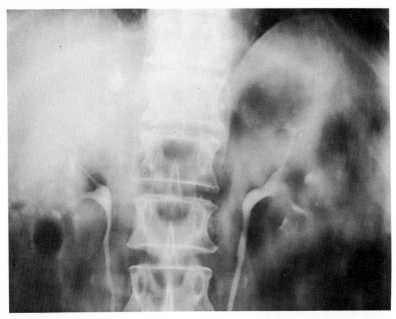
(2)

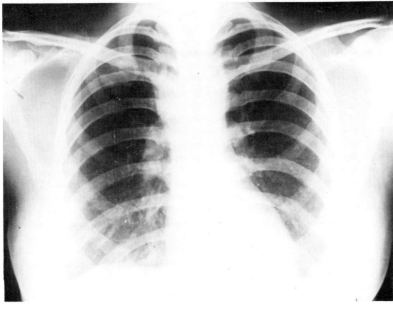
(3)

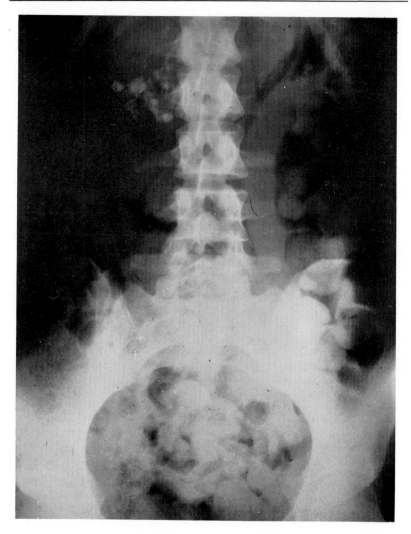

Q5.11

(a) Name the abnormal radiological sign.
(b) Name three ways in which this patient may present.
(c) What is the most likely cause of this patient's disease?

ANSWERS PAGE (226)

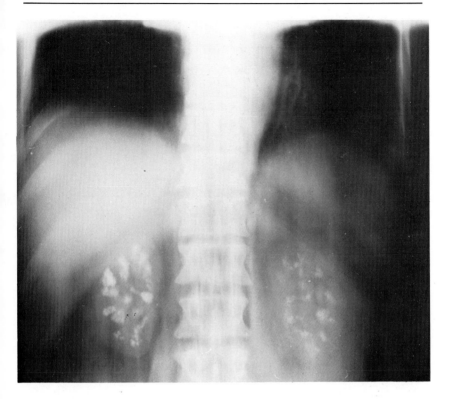

Q5.12 This disorder was inherited. The patient presented with osteomalacia.

(a) What is the most likely diagnosis?
(b) What special test would you do to confirm your diagnosis?

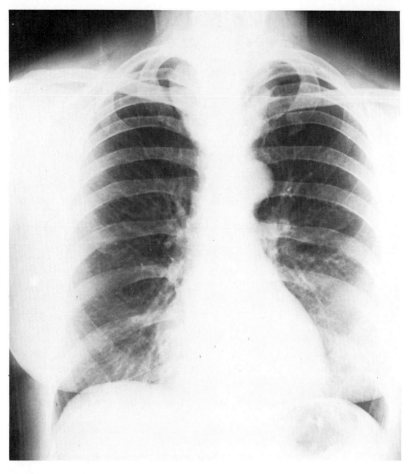

Q5.13 This patient presented with cardiac failure.
(a) Describe the abnormal radiological sign.
(b) What is the most likely cause of this radiological sign?
(c) Name three alternative causes of this radiological sign.
(d) Why was the patient in cardiac failure?

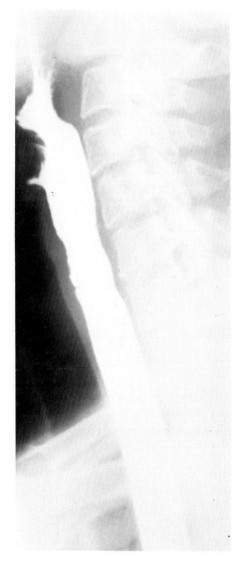

Q5.14 This lady is anaemic.
  (a) What is the radiological sign visible on this barium swallow?
  (b) What is the most likely diagnosis?

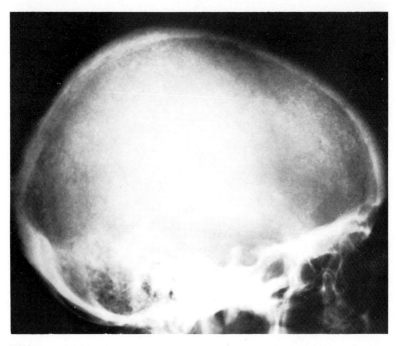

(1)

Q5.15

(a) Name the biochemical abnormality which is common to these three patients.
(b) What is the diagnosis in each case?

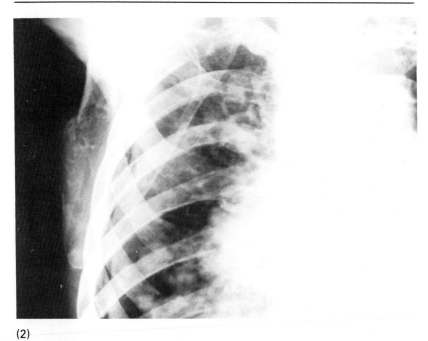
(2)

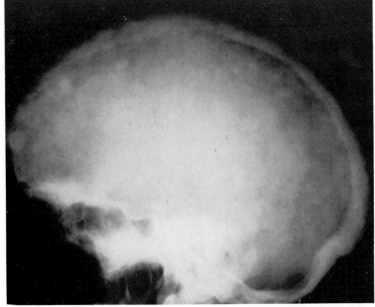
(3)

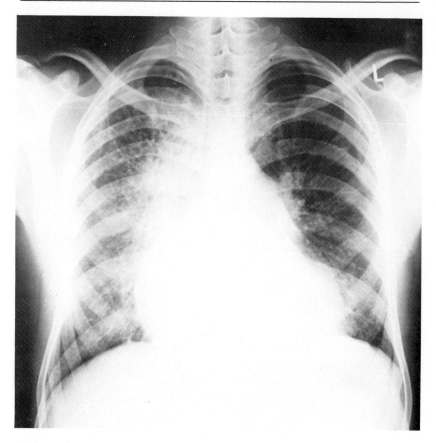

Q5.16 This 19 year old Iraqi has tuberculosis.
(a) From what other disorder is he suffering?
(b) What abnormality may be found on the electrocardiogram?

ANSWERS PAGE (231)

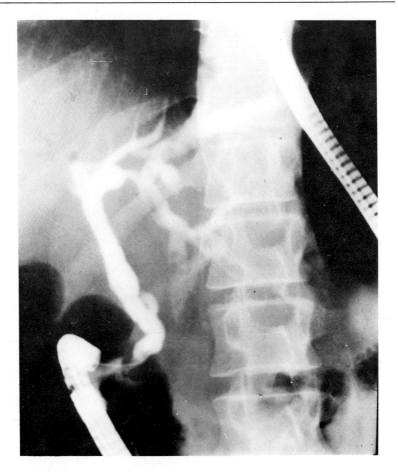

Q5.17 This 50 year old Japanese man presented with cholangitis.

(a) Name the radiological signs visible on this endoscopic retrograde cholangiogram.
(b) What is the most likely diagnosis?

ANSWERS PAGE (231)

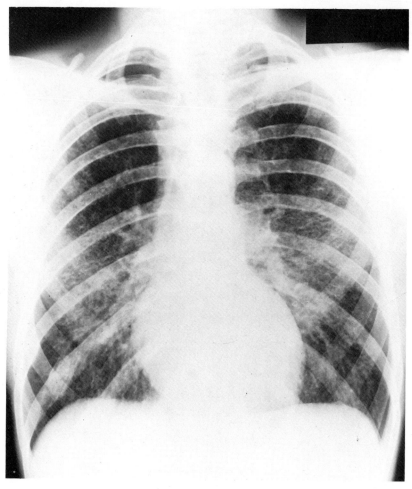

Q5.18 This young man presented with polydipsia and polyuria. He had prominent eyes, seborrheic dermatitis, and necrotizing skin lesions on face and in axillae.

(a) What is the most likely diagnosis?
(b) Name three other manifestations of this disease.

ANSWERS PAGE (232)

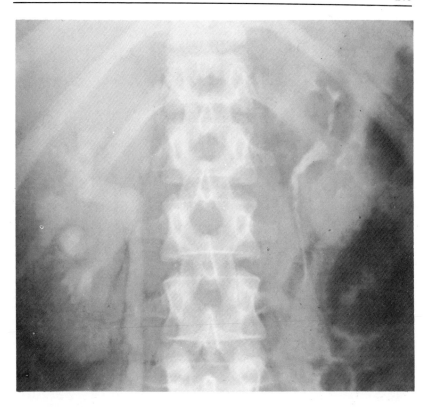

Q5.19 This 30 year old patient was asymptomatic. This film was taken 15 minutes after injection of contrast medium. The right kidney was not obstructed.
  (a) What is the diagnosis?
  (b) What is the most likely cause of this disorder in this patient?

ANSWERS PAGE (233)

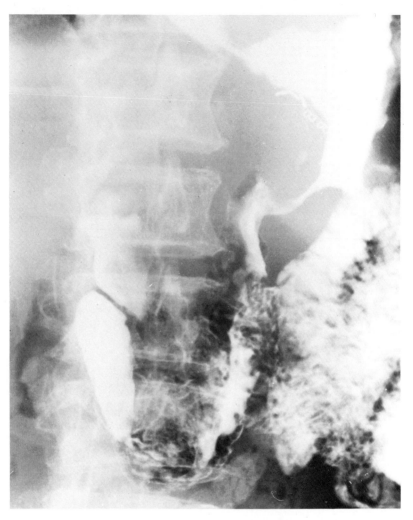

Q5.20 This patient presented with weight loss.
  (a) Name two possible causes of weight loss visible on this barium study.
  (c) Give three possible causes of anaemia in this patient.

ANSWERS PAGE (233)

PAPER 5

ANSWER

A5.1 (a) 1. Space occupying lesion in lower pole of left kidney with rotation of the kidney;
2. hydronephrosis of right ureter and kidney;
3. space occupying lesion in right side of bladder.
(b) Hypernephroma in left kidney. Bladder tumour at the right ureteric orifice.
(c) 1. Polycythaemia;
2. leukaemoid reaction:
3. plasmacytosis.

Only 55 per cent of all hypernephromas are symptomatic, the remainder being discovered inadvertently. About one third of all patients with this tumour present with non-urologic symptoms. Fever of unknown origin is present in nearly 25 per cent of all patients. Other presenting symptoms include weight loss, hypertension, anaemia, polycythaemia, hypercalcaemia and hepatic dysfunction.

Polycythaemia is an erythrocytosis, and differs from polycythaemia rubra vera in that splenomegaly, leukocytosis, and thrombocytosis are absent. A leukaemoid reaction with leukocytes ranging up to 100 000 per mm^3 and plasmacytosis have also been described.

A5.2 (a) 1. Dextrocardia, with situs inversus (gastric air bubble on right);
2. linear markings at right base.
(b) Bronchiectasis.

The labelling for right (R) shows that the X-ray should be reversed making dextrocardia and situs inversus obvious. Bronchiectasis may be congenital and associated with dextrocardia and sinusitis or absent frontal sinuses (Kartagener's syndrome). About a fifth of patients with dextrocardia have bronchiectasis. Bronchiectasis may also be associated with other congenital anomalies such as congenital heart disease, congenital kyphoscoliosis, and unilateral absence of pulmonary artery. However in the great majority of patients with bronchiectasis it is acquired, usually in childhood, as a result of bronchial infection and obstruction. Cough and purulent sputum are the characteristic symptoms and persistent crepitations in the same area of the chest are the most important physical signs of bronchiectasis. The bronchogram in this patient demonstrates saccular bronchiectasis of the left lower lobe.

A5.3 (a) **Choledocholithiasis.**
 (b) 1. **Biliary colic;**
     2. **acute cholangitis;**
     3. **pancreatitis;**
     4. **jaundice;**
     5. **secondary biliary cirrhosis.**

The percutaneous transhepatic cholangiogram shows a dilated common bile duct, packed with gallstones, and dilated hepatic ducts. Gallstones usually pass from the gallbladder into the common bile duct, but on rare occasions they form in the intrahepatic or common bile ducts. The stones may remain free in the duct, or obstruct it partially, completely or intermittently.

Choledocholithiasis may present without symptoms. Biliary colic occurs because of a sudden interference in bile flow which increases biliary pressure. The time interval between common duct obstruction and the onset of jaundice varies from 24 hr to several days depending on the distensibility and absorptive capacity of the gall bladder, the distensibility of the bile ducts and the rate of bile flow. When the intraductal pressure exceeds 25 cm H_2O hepatic bile flow is suppressed and conjugated bilirubinaemia occurs.

Acute cholangitis results in Charcot's triad — spiking fever with shaking chills, biliary colic and jaundice. The bile in three quarters of patients with acute cholangitis contains bacteria. Complete obstruction of the common duct may consequently produce acute suppurative cholangitis. This is a surgical emergency because, without surgical relief, the mortality approaches 100 per cent. Pancreatitis may accompany common duct obstruction presumably because the stone interferes with the outflow of pancreatic secretions. Continued or intermittent common duct obstruction may lead to secondary biliary cirrhosis with pruritus, steatorrhoea and loss of fat soluble vitamins.

A5.4 1. **Hydropneumothorax.**
2. **Pulmonary effusion with raised left hemidiaphragm.**
3. **Right pleural effusion with right paratracheal node enlargement and right cervical rib.**

A trace of fluid at the costophrenic angle, giving a small fluid level on chest X-ray, is a frequent complication of pneumothorax. A large quantity of fluid (clear exudate, blood or pus) may accumulate within the pleural cavity to produce the characteristic radiological picture of a hydropneumothorax. A horizontal fluid level separates the air above from the fluid below.

Pleural fluid which is predominantly along the diaphragmatic border may mimic an elevated hemidiaphragm. The easiest way to prove this is to do a decubitus film during which the patient should lie on his side. If a sub-pulmonary effusion is present the fluid will track along the costal margin and the diaphragm will become visible (Fig. A5.4).

The third patient had tuberculosis, with adenopathy and effusion. The cervical rib was asymptomatic.

Cervical rib is a common anomaly which usually arises from the seventh cervical vertebra. Rarely it arises from the sixth or even the fifth cervical vertebra. Cervical ribs vary greatly in size and shape. The symptoms, resulting from compression of the lower part of the brachial plexus or compression of the subclavian artery, have little relationship to the size of the radiographic abnormality. A small rib may have a fibrous attachment which causes much disability, whereas a large cervical rib may be asymptomatic.

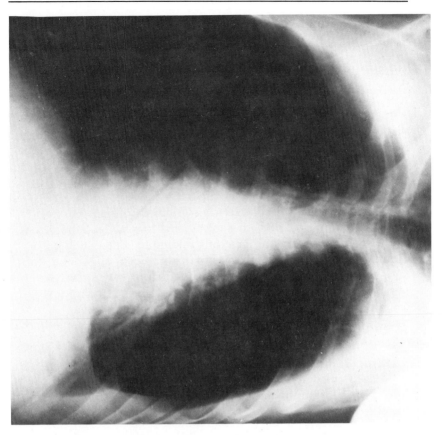

Fig. A5.4 Left lateral decubitus, chest X–ray showing how the fluid from a sub-pulmonary effusion in Patient 2 moves with change of position.

A5.5 (a) Bolus of food (acting as a ball valve) at the proximal end of benign oesophageal stricture.
(b) 1. Blood loss;
2. oesophagitis with heartburn or dysphagia;
3. weight loss.

This patient has a benign oesophageal stricture as shown by the distal oesophageal narrowing without shouldering or irregularity of the mucosa. Both severe oesophagitis and peptic ulcer of the oesophagus lead to peptic stricture.

Reflux oesophagitis is inflammation of the oesophageal mucosa caused by reflux of acid gastric contents or alkaline intestinal juice. This results in the symptoms of heartburn and dysphagia with hot, cold, or slightly acid foods. Oesophagitis is the precursor of the complications related to reflux such as blood loss, stricture and change in oesophageal epithelium from squamous to columnar. Columnar epithelium may show metaplastic changes which are believed to be premalignant. Development of a benign stricture leads to progressive dysphagia and, with advancing disease, malnutrition is inevitable.

A5.6 (a) Brain abscess.
(b) Penicillin 20 million units I.V per day for 6 weeks. Chloramphenicol 4-6 g I.V for 6 weeks.

The post contrast CT scan shows a lesion in the right frontal lobe with ring enhancement (surrounding low attenuation area) and a mass effect. The appearance of a ring flush is a nonspecific indication of pathology and seen frequently with gliomas, metastases and abscesses, and occasionally in patients with aneurysm, infarction or haematoma. The ring flush may represent breakdown of the blood brain barrier. In patients with brain abscess CT scan is very useful because it permits (a) demonstration of the presence, site and number of abscesses; (b) documentation of capsule formation and/or impending capsule rupture (as a guide to timing of operative intervention); (c) and aspiration of abscesses with CT guidance.

Nearly half of all brain abscesses are secondary to disease of the middle ear and mastoid cells, and occur in the cerebellum or temporal lobe. Frontal sinusitis accounts for about 10 per cent of cases, the abscess being usually situated in the frontal lobe.

Of the remaining cases, a small proportion are due to penetrating wounds or post operative infections and the rest are metastatic from a primary septic focus, usually in the lung but also in skin, bone, or heart. Brain abscesses are particularly frequent in congenital heart disease with right-to-left shunts.

Many brain abscesses can be cured by administration of adequate doses of antibiotics. The most common organisms causing abscesses are streptococci, most of which are anaerobic or microaerophilic, in combination with other anaerobes (such as bacteroides) or enterbacteriacae (such as E. coli, Klebsiella or Proteus). Staphylococci may also cause brain abscess, usually as a consequence of penetrating head trauma or bacteraemia. As the bacteriological diagnosis is presumptive one recommended regimen for brain abscess is parenteral penicillin and chloramphenicol in high doses for six weeks. Metronidazole may also be added to this regimen to cover anaerobic organisms. Failure to improve with antibiotic treatment or progression of high intracranial pressure are indications for surgery.

A5.7 (a) **Terminal phalangeal erosion.**
  (b) 1. **Mediastinal opacities,**
     2. **Basal shadowing with decreased lung volume.**
  (c) **Scleroderma**

In scleroderma, the taut skin over the skin limits full extension and fixed flexion deformities develop. The soft tissue of the fingertips is lost and occasionally the bone of the terminal phalanges is resorbed. The first symptoms are stiffness of the fingers and Raynaud's phenomenon. The latter symptom may precede the skin changes by months or even years. In severe cases, bilateral cervical sympathectomy has been used to ease the symptoms of Raynaud's phenomenon. The mediastinal opacities in this patient resulted from bilateral cervical chain sympathectomies.

Pulmonary insufficiency due to fibrosis accounts for about one fifth of the deaths in scleroderma. These patients often complain of a dry cough or exertional dyspnoea. Even in the early stages, they may have low diffusion capacity and low PO_2 on exercise. The radiological changes of linear densities, mottling and, finally, honeycombing are more evident in the lower two thirds of the lungs in scleroderma.

A5.8 (a) Perforation of the gut.
(b) Zollinger-Ellison syndrome.
(c) Serum gastrin level.

This plain abdominal film shows air outlining the outside of the gut, giving a double-contrast effect with the air inside the gut. Perforation of the gut is an important complication of peptic disease, malignant gut disease and diverticular disease. Some 90 to 95 per cent of patients with gastrinomas develop ulceration of the gastrointestinal tract at some point during the course of their disease and are especially prone to gut perforation. Profound acid hypersecretion induces fulminant, progressive and persistent ulcer disease. Multiple ulcers may occur not only in the stomach and in the first part of the duodenum but also in the remainder of the duodenum or even jejunum. Diarrhoea occurs in about 40 per cent of patients. It is due to the large amounts of hydrochloric acid in the proximal duodenum, with consequent lowering of gut pH; to inflammatory changes in the small gut mucosa; and to steatorrhoea. Inactivation of pancreatic lipase and precipitation of bile acids by the lowered pH in proximal small intestine causes steatorrhoea and malabsorption.

Ninety per cent of gastrinomas occur in the pancreas. About two thirds of them are malignant but patients usually die from fulminant peptic disease rather than metastases. The diagnosis of Zollinger-Ellison syndrome depends upon the demonstration of increased serum gastrin levels, which should be obtained in patients with a history of severe peptic disease and marked acid hypersecretion.

A5.9 (a) Breathlessness
(b) 1. Smoking;
2. air pollution;
3. α_1 antitrypsin.
(c) 1. Reduced Forced Vital Capacity (FVC)
2. reduced Forced Expiratory Volume (FEV$_1$);
3. increased Residual Volume;
4. normal or reduced PO$_2$;
5. normal PCO$_2$.

Distension of the lung in this patient with emphysema resulted in low flattened diaphragms with a long thin cardiac silhouette. Enlargement of the proximal parts of the pulmonary arteries in the hila occurs if hypoxaemia develops and produces pulmonary hypertension.

α_1 Antitrypsin deficiency probably only accounts for a very small proportion of patients with emphysema and in these it is aggravated by cigarette smoking and air pollution. The deficiency should be particularly sought in patients developing dyspnoea and emphysema under the age of 40, especially if they are women or non-smokers.

In patients with chronic obstructive airways disease in whom the emphysema component is dominant the major disability is breathlessness. These patients are the 'pink puffers' who are able to maintain their PaO_2 and PCO_2 (up to a late stage of the disorder) by hyperventilation. Therefore pulmonary hypertension and consequently cor pulmonale does not usually develop, and respiratory failure is the usual cause of death. Over-inflation of the lungs may be demonstrated by the increased residual volume with reduced forced vital capacity. Most patients with emphysema have reduced FEV_1 and peak expiratory flow rate, suggesting an element of airways obstruction.

A5.10 (a) Tuberculosis
   (b) 1. Ileal stricture with splenic calcification;
   2. right adrenal calcification;
   3. right paratracheal node enlargement with right apical shadow.
   (c) Fever, right lower quadrant pain, diarrhoea.

The onset of tuberculosis is usually insidious and the patient may be entirely asymptomatic. Many cases are discovered during routine chest X-rays. The earliest symptoms are malaise, excessive fatigue, night sweats and weight loss as occurred in Patient 3.

The normal gastrointestinal tract is resistant to penetration by tubercle bacilli but when large numbers of bacilli are present the mucosa may be penetrated in the ileo-caecal region. This causes symptoms of intermittent abdominal pain, diarrhoea and weight loss. Radiologically stenotic lesions are visible in the terminal ileum. Haematogenous tuberculosis may sometimes localise in the adrenal glands. Total destruction of the adrenals gives rise to Addison's disease. However Addisons' disease is unlikely to occur in patients with unilateral adrenal calcification.

A5.11 (a) **Pancreatic calcification**
    (b) 1. **Abdominal pain;**
        2. **malabsorption;**
        3. **diabetes mellitus.**
    (c) **Alcohol**

The radiographic hall mark of chronic pancreatitis is the presence of scattered calcification throughout the pancreas. This indicates that significant damage has occurred in the pancreas and therefore may be associated with steatorrhoea and diabetes mellitus.

Exocrine pancreatic insufficiency is found in less than one-third of chronic pancreatitis patients. Abdominal pain is characteristically epigastric radiating through to the back, but it may be maximal in the right or left upper quadrants, in the back, or diffuse throughout the upper abdomen, or even referred to the anterior chest or flank. The pain is usually persistent, deep seated, unresponsive to antacids and often increased by alcohol and heavy meals. In addition it may be so severe and continuous as to require use of narcotics, or it may be intermittent or even absent. Alcohol is by far the most frequent cause of pancreatic calcification, but may be seen in severe protein-calorie malnutrition, hyperparathyroidism, hereditary pancreatitis, post traumatic pancreatitis and islet-cell tumours.

A5.12 (a) **Distal renal tubular acidosis.**
    (b) **Oral ammonium chloride test.**

Nephrocalcinosis is the term applied to multiple papillary calcifications found on X–ray. It is very common in hereditary distal renal tubular acidosis (Type 1), but also occurs in other states characterised by severe hypercalciuria, such as idiopathic hypercalciuria, primary hyperparathyroidism, and sarcoidosis.

Distal renal tubular acidosis is usually of autosomal dominant inheritance. Hyperchloraemic acidosis associated with alkaline urine occurs because the kidney does not lower urine pH normally, thus resulting in a deficient excretion of acid. Urinary osmotic concentration and potassium retention is also impaired, causing polyuria and hypoklaemia. The chronic acidosis lowers tubule reabsorption of calcium, causes hypercalciuria and secondary hyperparathyroidism. The result is calcium phosphate stones or nephrocalcinosis. Bone disease (rickets in children or osteomalacia in adults) occurs because of acidosis-induced loss of bone mineral and inadequate production of 1,25 dihydroxy-vitamin D_3.

The diagnosis of distal renal tubular acidosis can be confirmed by giving 0.1 g of ammonium chloride orally and following the blood and urine pH. In affected patients, urine pH will not fall below 5.5, although systemic acidosis worsens.

A5.13 (a) **Superior mediastinal mass with tracheal deviation.**
(b) **Goitre**
(c) **Thymic mass, lymphadenopathy, innominate artery aneurysm.**
(d) **Thyrotoxic atrial fibrillation.**

The commonest abnormal mediastinal mass is retrosternal goitre. It is most frequent in middle aged females who may be thyrotoxic. In general nervous symptoms dominate the clinical picture in young thyrotoxic individuals whereas cardiovascular and myopathic symptoms predominate in older subjects. The posteroanterior chest X-ray shows superior mediastinal shadow which looks like an inverted truncated cone. The trachea is characteristically displaced posteriorly but on the P.A. film a lateral shift may be observed. Another important radiological sign of retrosternal goitre is nodular calcification, not present in this patient.

A5.14 (a) **Post-cricoid oesophageal web.**
(b) **Plummer-Vinson syndrome.**

Iron deficiency anaemia in middle-aged females may be associated with dysphagia and a radiologically demonstrable web in the post cricoid region. These webs may occur without anaemia, although iron deficiency may be present. They sometimes precede the development of post cricoid carcinoma. The webs appear as thin filling defects arising from the anterior wall of the cervical oesophagus immediately below the cricopharyngeus, and are seen only when the oesophagus is distended with barium.

A5.15 (a) **Hypercalaemia**
  (b) 1. **Hyperparathyroidism;**
      2. **lytic metastasis to scapula**
      3. **Paget's disease**

Hypercalcaemia with low serum phosphate and hyperchloraemic acidosis suggests primary hyperparathyroidism. The skull X-ray has a characteristic appearance, often described as 'miliary osteoporosis' or 'pepper pot porosis'.

Malignant processes may produce hypercalcaemia, through several different mechanisms, including dissolution of bone from metastases, and ectopic parathormone production. As a rule osteolytic metastases produce hypercalcaemia, and commonly arise from lung, thyroid, kidney and lower bowel. Osteoblastic metastases are usually radiodense and accompanied by a rise in alkaline phosphatase and normal, or even low, calcium levels, (e.g. prostatic carcinoma). However, some tumours, (especially carcinoma of the breast), produce both osteolytic and osteoblastic secondaries. Then there may be phases in which osteolysis predominates, alternating with phases in which alkaline phosphatase levels rise and the skeletal lesions become more sclerotic.

In Paget's disease, the radiological signs are also characteristic. An increase in thickness of the skull vault with an irregular mottled woolly appearance may be seen. A large increase in cortical width may be found in the femur or tibia, (Fig. A5.15), and these bones may be very bowed anterolaterally. Four types of bone change may be seen in Paget's disease —(i) lytic active areas, (ii) new bone growing from cortex into medulla with coarse irregular trabeculations, (iii) diffuse, dense, enlarged, inactive areas, and (iv) sarcomatous change.

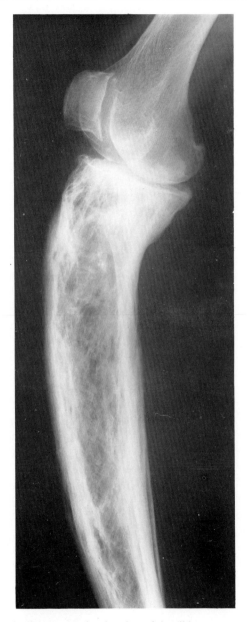

Fig. A5.15 Paget's disease causing bowing of the tibia.

A5.16 (a) Mitral stenosis.
(b) P-Mitrale.

The chest X-ray of this patient shows widespread mottling in the right upper zone, with small cavities and slight shrinkage, due to tuberculosis. In addition the heart is grossly enlarged, with double contour to the right heart border and prominent left atrial appendage, due to mitral stenosis.

The earliest radiographic changes in mitral stenosis are straightening of the left border of the heart silhouette, prominence of the main pulmonary arteries, dilation of the upper lobe pulmonary veins and backward displacement of the oesophagus by an enlarged left atrium (Fig. A5.16). Usually the overall cardiac size is grossly enlarged only in severe mitral stenosis, when both atria, pulmonary arteries and veins, right ventricle and superior vena cava are prominent. Kerley B lines, when present, signify a mean atrial pressure of at least 20 mm Hg. As the pulmonary arterial pressure rises, the smaller pulmonary arteries become attenuated at first in the lower, then in the mid, and finally in the upper lung fields. Deposits of haemosiderin occur in the lungs of patients who have had multiple haemoptyses. This may result in fine diffuse nodules, most prominent in the lower lung fields, which may ossify.

The classical P-mitrale consists of a bifid P-wave measuring 0.10 seconds or more in width with normal or slightly increased amplitude. The first peak is due to the right atrium and the second to the left atrium. Occasionally instead of a bifid P-wave it is tall with a narrow base (P-pulmonale). This is due to a high pulmonary vascular resistance, in which case the electrocardiogram also shows right ventricular hypertrophy.

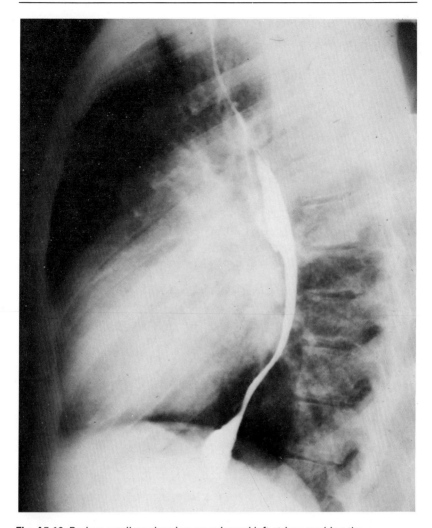

Fig. A5.16 Barium swallow showing an enlarged left atrium pushing the oesophagus posteriorly.

A5.17 (a) 1. Stenosed right main hepatic duct;
2. stricture in left main hepatic duct with distal dilation.
(b) **Oriental cholangiohepatitis.**

In the Orient, many persons of low socioeconomic status are affected by oriental cholangiohepatitis. This is a form of chronic recurrent cholangitis caused by calcium bilirubin stones formed within bile ducts, perhaps resulting from parasites in the ducts. Acute cholangitis is the usual presentation. Strictures sometimes develop in the hepatic ducts, leading to further stasis and intrahepatic stone formation. Most acute cases of cholangitis can be controlled with antibiotics but severe toxicity or septic shock requires emergency surgery. If nonsurgical management of the acute disease is successful, transhepatic or endoscopic retrograde cholangiograms may be used to outline the ducts to discover the cause of cholangitis.

A5.18 (a) **Multifocal eosinophilic granuloma with pulmonary fibrosis.**
(b) 1. Osteolytic lesions;
2. adenopathy;
3. hepatosplenomegaly;
4. mastoiditis.

This patient's chest X–ray shows shadowing in middle and lower zones with cystic changes suggestive of honeycomb lung. Classically multifocal eosinophilic granuloma comprises skull lesions, exophthalmos and diabetes insipidus, a triad which occurs in about a quarter of the patients with this disease. Nonspecific symptoms such as malaise, anorexia, fever and irritability are common, as is mastoiditis with accompanying upper respiratory tract infections and otitis media. Lymphadenopathy and hepatosplenomegaly occur in about a quarter of patients and pulmonary fibrosis is a well recognized complication. Osteolytic bone lesions may be multiple and are found predominantly in flat bones.

Traditionally eosinophilic granuloma, Hand-Schuller-Christian disease and Letterer-Siwe disease have been considered as a single group and designated 'Histiocytosis X'. Now Letterer-Siwe is regarded as a lymphomatous proliferation of poorly differentiated histiocytes, whereas eosinophilic granuloma and Hand-Schuller-Christian disease are thought to represent a nonneoplastic reaction of well differentiated histiocytes to an unknown stimulus and can be referred to as unifocal and multifocal eosinophilic granuloma respectively.

A5.19 (a) Left renal artery stenosis.
(b) Fibromuscular hyperplasia.

The onset of hypertension under 30 years of age or after 55 years of age makes an intravenous pyelogram imperative. Renal artery stenosis is usually caused by fibromuscular hyperplasia in the younger age group and atherosclerosis in the older group. If there is unilateral occlusion of the renal artery the contrast medium appears first in the pelvis and ureter of the unaffected kidney whereas excretion of the contrast medium on the affected side is delayed. However ten minutes or so later the picture is that of a dense, small volume, pyelogram on the affected side. The intravenous pyelogram in this patient illustrates a small left kidney, a large right kidney, and a denser pyelogram with smaller calyceal volume on the left side compared to the right. There is usually nothing particularly striking about the presentation of patients with renal artery stenosis which distinguishes them from others with essential hypertension, unless there has been an episode of loin pain with haematuria, pointing to a renal infarct.

A5.20 (a) 1. Gastric carcinoma;
 2. blind loop syndrome.
 (b) 1. B_{12} deficiency;
 2. blood loss;
 3. folic acid deficiency.

The barium study demonstrates a Billroth II partial (Polya) gastrectomy (surgical clips visible) with narrowing and shouldering in the residual antrum suggestive of gastric carcinoma. Among the long term complications following partial gastrectomy are gastric carcinoma, malabsorption, pulmonary tuberculosis and anaemia. Multiple haematinic deficiency may occur.
 Following partial gastrectomy chronic blood loss occurs with such frequency that oral iron is frequently prescribed to prevent the development of iron deficiency. In addition, malabsorption of

iron in the jejunum may also occur. In some patients who have had extensive gastric surgery or have mucosal damage, the source of intrinsic factor may be absent and B_{12} deficiency develops. In addition colonization of blind loop and small intestine with bacteria may cause further B_{12} deficiency because the bacteria utilize B_{12}. Folic acid deficiency occurs in a small proportion of patients who have had partial gastrectomy, because they are likely to eat less than normal and because malabsorption, particularly following a Billroth II anastomosis, may occur. In addition to bacterial overgrowth malabsorption develops because there is a decreased pancreatic enzyme response to food when the duodenum is bypassed; inadequate mixing of pancreatic enzymes occurs; rapid emptying of the stomach and rapid transit through the small gut is present.

REFERENCES

Medicine

Beeson P B et al (Ed) 1979 Cecil's Textbook of Medicine Saunders, Philadelphia 15th Edn
Isselbacher K J et al (Ed) 1980 Harrison's Principles and practice of internal medicine McGraw-Hill, New York 9th Edn
Medicine. The monthly add-on journal, 2nd and 3rd series. Medical Education (International) Ltd, Oxford
The Medical Clinics of North America. Published every other month by Saunders, Philadelphia

Radiology

Kreel L 1971 Outline of radiology. Heinemann, London.
Sutton D 1980 A textbook of radiology and imaging. Churchill Livingstone, Edinburgh & London.

INDEX

Cardiology

Aortic incompetence, 149
Aneurysm
  ascending aorta, 16, 17
  descending aorta, 162
  dissecting, 23
Atrial septal defect
  pre-surgery, 116
  post-surgery, 138
  Eisenmenger's syndrome, 56
Dextrocardia, 194
Fallot's tetralogy, 66
Mitral stenosis, 212, 231
Pericardial calcification, 99, 126
Pericardial effusion, 105
Superior vena caval obstruction, 112
Total anomalous pulmonary venous drainage, 150
Ventricular aneurysm, 146

Endocrinology

Acromegaly
  skull, 144
  heel pad, 145
  hands, 171
Adrenal calcification, 205
Goitre, retrosternal, 208
Hyperparathyroidism
  bone cyst, 52
  subperiosteal erosions, 102
  pepper pot skull, 210
  Rugger jersey spine, 161
Osteomalacia, 121
Osteoporosis, 59
Pheochromocytoma, 69, 82
Thyroid carcinoma, 12

Gastroenterology

Achalasia, 14
Amoebic abscess, 73
Amyloid of small gut, 148
Ascariasis, 64
Blind loop 216
Cholangiohepatitis, 213
Choledochoduodenal fistula, 50
Choledocholithias, 194
Crohn's disease, 156
Diverticulosis, oesophageal candidiasis, 58
Fistula, 118
Gardener's syndrome, 72
Gastric carcinoma, 34, 216
Ileal stricture, 204
Ileus, 51
Intussusception, 29
Jejunal diverticulosis, 50, 117
Oesophageal stricture, 198
Oesophageal varices, 18
Oesophageal web, 209
Pancreatic calcification, 206
Partial gastrectomy, 216
Perforation, 202
Peritoneal metastases, 108
Pyloric stenosis, 147
Ulcerative colitis, 110, 134

Haematology

Lymphoma
  abdominal, 54
  mediastinal, 104
Multiple myeloma, 163
Plummer-Vinson syndrome, 209
Thalassaemia major, 74, 96

Infectious diseases

Amoebic abscess, 73
Ascariasis, 64
Aspergilloma, 8, 36
Candidiasis of oesophagus, 58
Gas gangrene, 155
Guinea worm, 71, 93
Histoplasmosis, 67
Pneumococcal pneumonia, 65, 86
Staphylococcal pneumonia, 51, 77
Syphilis
  aneurysm of aorta, 16, 17
  bismuth injection, 100
  charcot joint, 101
Tuberculosis
  adrenal, 205
  apical, 157, 205, 212

Tuberculosis (cont'd)
  ileal, 204
  mediastinal, 205
  miliary, 30, 31
  pericardial, 99, 126
  plombage, 98
  primary complex, 98, 127
  thoracoplasty, 99

Nephrology

Hypernephroma, 102, 192
Medullary sponge kidney, 10
Nephrocalcinosis, 207
Papillary necrosis, 114, 115
Renal artery stenosis, 215
Renal stones, 131
Retroperitoneal fibrosis, 68
Rugger jersey spine, 161
Staghorn calculi 25
Ureteric duplication, 106
Ureteric obstruction, 21, 54, 193
Urothelial tumour, 103

Neurology

Cerebral abscess, 199
Cerebral metastases, 12
Charcot joint, 101
Hydrocephalus, 153
Infarct of internal capsule, 26
Intervertebral lumbar disc lesion, 164
Intracranial aneurysms, 62, 63
Medulloblastoma, 152
Meningioma, 19
Spinal cord compression 82
Subdural haematoma, 107

Respiratory disease

Aspergilloma, 8, 36
Azygos vein and lobe, 70
Bronchial foreign body, 160
Bronchiectasis, 32, 194
Bulla, 109
Cervical rib, 197
Collapse
  left lung, 154
  left lower lobe, 87, 88
Consolidation
  left lower lobe, 65, 86
  right lower lobe, 158, 179
  right middle lobe, 180
  apical segment right lower lobe, 166, 186
  lingula, 168, 186
  posterior segment right upper lobe, 167, 187

Emphysema, 109, 203
Hilar lymphadenopathy, 20
Hydropneumothorax, 196
Hypertrophic osteoarthropathy, 159
Lymphangitis carcinomatosa, 11
Mediastinal emphysema, 24
Mediastinal lymphadenopathy, 104, 205
Metastases from thyroid carcinoma, 13
Miliary tuberculosis, 30, 31
Pancoast's tumour, 119
Pleural effusion, 123, 197, 221
Pleural tumour, 123
Plombage, 98
Pneumonia
  air bronchogram, 89
  pneumococcal, 65, 86
  staphylococcal, 51, 77
Pneumothorax, 109
Primary complex, 88, 127
Progressive massive fibrosis, 113
Pulmonary eosinophilia, 61
Pulmonary fibrosis, 149, 201, 214
Thoracoplasty, 99

Rheumatology

Avascular necrosis, 27, 60
Calcinosis, 163
Chondrocalcinosis, 168
Gout, 33
Osteoarthrosis, 15
Paget's disease, 211, 229
Polyarteritis nodosa, 9
Psoriatic arthritis, 111
Scleroderma, 200, 201

General

Allbright's syndrome, 151
Cleido-cranial dysostosis 70, 93
Collapsed vertebra, 59
Dermoid cyst, 120
Fibroid, 120
Gaucher's disease
  avascular necrosis, 27
  splenomegaly, 26
Neurofibromatosis, 57, 82
Osteoblastic metastasis, 55
Osteolytic metastasis, 211
Ovarian cyst, 53, 79
Sarcoidosis
  bone cyst, 21
  hilar lymphadenopathy, 20